BIOTECHNOLOGY AND THE ENVIRONMENT

BIOTECHNOLOGY AND THE ENVIRONMENT

Research Needs

Edited by

Gilbert S. Omenn, M.D., Ph.D., Dean
School of Public Health and Community Medicine
University of Washington
Seattle, Washington

and

Albert H. Teich, Head
Office of Public Sector Programs
American Association for the Advancement of Science
Washington, DC

NOYES DATA CORPORATION
Park Ridge, New Jersey, U.S.A.
1986

363.7
B616

Copyright © 1986 by Noyes Data Corporation
Library of Congress Catalog Card Number 86-18181
ISBN: 0-8155-1105-1
Printed in the United States

Published in the United States of America by
Noyes Data Corporation
Mill Road, Park Ridge, New Jersey 07656

10 9 8 7 6 5 4 3 2 1

Library of Congress Cataloging-in-Publication Data

Biotechnology and the environment.

    Includes bibliographies and index.
    1. Biotechnology--Environmental aspects--Congresses.
I. Omenn, Gilbert S. II. Teich, Albert H.
TP 248.13.B57   363.7'01   86-18181
ISBN 0-8155-1105-1

Foreword

This book discusses research needs relating to the effects of biotechnology on the environment. The material presented is based on, and the result of, a workshop convened by the American Association for the Advancement of Science and the U.S. Environmental Protection Agency at Berkeley Springs, West Virginia in 1984.

The manner in which the release of genetically-altered organisms will effect the environment, and indeed whether they should be released at all, has been a subject of great interest for many years. That it has become a highly controversial issue recently is only a small indication of the enormous advances being made almost daily in the field of biotechnology. Many questions must be answered and numerous problems solved—among them how to monitor and control the environmental effects of the release or synthesis of novel organisms. This should be a valuable document for those concerned with the release of genetically-altered organisms to the working or general environment.

The book presents a workshop summary and four papers which cover major areas of concern—environmental effects, health effects, monitoring and quality assurance, and control technologies—as each of these applies to research needs in biotechnology and the environment.

The information in the book is from *Research Needs in Biotechnology and the Environment,* edited by Gilbert S. Omenn of the University of Washington School of Public Health and Community Medicine and Albert H. Teich of the American Association for the Advancement of Science, Office of Public Sector Programs, prepared for the U.S. Environmental Protection Agency by the American Association for the Advancement of Science, November 1985.

vi Foreword

The table of contents is organized in such a way as to serve as a subject index and provides easy access to the information contained in the book.

> Advanced composition and production methods developed by Noyes Data Corporation are employed to bring this durably bound book to you in a minimum of time. Special techniques are used to close the gap between "manuscript" and "completed book." In order to keep the price of the book to a reasonable level, it has been partially reproduced by photo-offset directly from the original report and the cost saving passed on to the reader. Due to this method of publishing, certain portions of the book may be less legible than desired.

ACKNOWLEDGMENTS

Many people contributed to the workshop and to this report and deserve thanks for their contributions. Among them are, of course, the paper authors, and other workshop participants, who gave generously of their time and whose names are listed in an appendix at the end of this report, and, especially, the workshop chairman, Gilbert S. Omenn. EPA officials, led by John R. Fowle III and Morris A. Levin, were most supportive and helpful in arranging and conducting the workshop and in preparing this report. We also appreciate the assistance of the consultants who prepared the two commissioned papers following the workshop, Marvin Rogul and David Glaser and his colleagues. Finally, many current and former AAAS staff members contributed their time and energy to the project in its various stages, including Jill H. Pace, Barbara Dworsky, and, especially, Mary I. Haddock.

NOTICE

The materials in this book were prepared as accounts of work sponsored by the American Association for the Advancement of Science and the U.S. Environmental Protection Agency. Publication does not signify that the contents necessarily reflect the views and policies of the contracting agencies or the Publisher, nor does mention of trade names or commercial products constitute endorsement or recommendation for use. To the best of our knowledge the information contained in this publication is accurate; however, the Publisher assumes no liability for errors or any consequences arising from the use of the information contained herein.

Contents and Subject Index

PART I
WORKSHOP SUMMARY

WORKSHOP SUMMARY 2
Gilbert S. Omenn and Albert H. Teich
   Introduction and Background 2
   Key Concerns and Recommendations 6
   Environmental Effects 9
   Health Effects 11
   Monitoring and Quality Assurance 13
   Control Technologies 14

PART II
PAPERS

RESEARCH PLAN FOR TEST METHODS DEVELOPMENT FOR RISK ASSESSMENT OF NOVEL MICROBES RELEASED INTO TERRESTRIAL AND AQUATIC ECOSYSTEMS 18
Al Bourquin and Ramon Seidler
   I. Introduction 18
      A. Goal 18
      B. Risk Assessment 19
      C. Two Major Approaches 20
         1. Data Base Development 20
         2. Research 21
      D. Short and Long Term Needs 21
      E. Relevance to EPA Needs 22
   II. **Novel Organisms** 24

III. **Development of Test Methods for the Detection, Identification, and Enumeration of Novel Organisms** 26
   A. Statement of Research Problems 26
   B. Availability of Data Base. 27
   C. Approaches 27
      1. Conventional Techniques 27
      2. Molecular Techniques. 29
   D. Short Term Products 31
   E. Long Term Products 31
IV. **Development of Test Methods for Assessing Fate of Novel Organisms** 33
   A. Statement of Research Problems 33
   B. Availability of Data Base. 34
   C. Approaches 35
      1. The Microcosm Approach 35
      2. Rationale for Selecting Ecosystems. 36
         a. Terrestrial Research 36
         b. Aquatic Research 37
   D. Short Term Products 39
   E. Long Term Products 40
V. **Development of Test Methods for Assessing Genetic Stability of Novel Organisms** 42
   A. Statement of Research Problems 42
   B. Availability of Data Base. 43
   C. Approaches 44
      1. Naked Plasmid DNA 44
      2. Stability of Plasmid DNA in Novel Organisms. 45
      3. Sources of Cultures and Plasmids 49
   D. Short Term Products 49
   E. Long Term Products 50
VI. **Development of Test Methods for Assessing Hazards of Released Novel Organisms.** 52
   A. Statement of Research Problems 52
   B. Availability of Data Base. 53
   C. Approaches 53
   D. Short Term Products 55
   E. Long Term Products 56
VII. **Summary.** 58
VIII. **Acknowledgments.** 60
IX. **Literature Cited** 61

BIOTECHNOLOGY HEALTH RISK ASSESSMENT RESEARCH PLAN. 68
Marvin Rogul and John R. Fowle III
   I. **Introduction.** 68
   II. **Health Effects Work Group Panel Recommendations** 70
      A. Data Gathering and Information Management. 70
      B. Selection of Organisms for Validating Subpart M Test Approach. 74

          C. Protocol Development for Infectivity, Pathogenicity, and
             Metabolic Characteristics of Recombinant Microorganisms ... 74
          D. Bacterial Pathogenicity Categories 75
          E. Establishment and Management of a Data Base of Charac-
             teristics of the Potential Hazards of Genetically Modified
             Materials 77
          F. Selection and Assessment of Safe Hosts 77
          G. Development of Molecular Probes................... 77
    III. Discussion 79
          A. Risk Assessment 79
          B. Foundation Laid by the NIH Recombinant DNA Advisory
             Committee 81
             1. *E. coli* Studies Which Influenced the Development of
                the RAC Guidelines............................ 82
             2. Experiments Simulating High Risk Conditions:
                Promoting and Detecting Genetic Interchange......... 83
    IV. References... 85

ENVIRONMENTAL ENGINEERING RESEARCH SUPPORT
PROPOSAL 86
John Burckle and Albert D. Venosa
    I.   Legislation... 86
    II. Regulatory Needs 86
    III. Overall Program Approach 89
    IV. **Summary of Proposed Environmental Engineering Efforts
         Related to Regulatory Needs**........................... 90
          A. Regulatory Needs 90
          B. Program Structure................................ 91
          C. Proposed Approach 92
          D. Development of Engineering Information and Methodology
             for Risk Assessment, Reduction and Management for
             Genetically Engineered Microorganisms in Biologically
             Based Manufacturing Processes and Deliberate
             Environmental Release 95
             1. Accidental and Deliberate Release from Biologically
                Based Manufacturing Processes
                1.1 Potential for Industrial Process Release 96
                1.2 Potential for Worker Exposure................ 99
                1.3 Technology for Containment (or Engineering
                     Controls) of Process Equipment Releases........ 101
                1.4 Identify Monitoring Needs and Strategies........ 103
                1.5 Decontamination Technology 104
                1.6 Worker Personal Protective Equipment 105
             2. Deliberate Release Into the Environment............ 107
                2.1 Site Profile Evaluation Procedure 107
                2.2 Site Containment Alternatives 109
                2.3 Monitoring Needs and Strategies.............. 110

　　　　　　　2.4　Alternative Site Decontamination Techniques
　　　　　　　　　　and Procedures..........................111
　　　　　　　2.5　Application Technology for Genetically
　　　　　　　　　　Engineered Microorganisms................113

**MONITORING TECHNIQUES FOR GENETICALLY ENGINEERED
MICROORGANISMS**..114
*David Glaser, Tim Keith, Peg Riley, Geoff Chambers, John Manning,
Susan Hattingh and Ralph Evans*
　　　I.　Introduction..114
　　　II.　Sampling Considerations................................117
　　　　　A. Introduction...117
　　　　　B. Qualitative Sampling.................................117
　　　　　C. Desorption from Sediments............................118
　　　　　D. Enrichment...118
　　　　　E. Partitioning in the Environment......................120
　　　　　F. Issues in Sampling Methods...........................121
　　　III.　Monitoring Techniques.................................123
　　　　　A. Conventional Microbiological Techniques..............123
　　　　　B. Immunological Techniques.............................124
　　　　　　　1. Standard Methods of Antibody Production..........124
　　　　　　　2. Monoclonal Methods of Antibody Production........127
　　　　　C. The Use of Genetic Markers...........................128
　　　　　D. Molecular Techniques.................................130
　　　　　　　1. Restriction Enzyme Mapping.......................130
　　　　　　　2. DNA Probes.......................................132
　　　　　　　　　a. Restriction Fragment Hybridization............133
　　　　　　　　　b. Colony Hybridization.........................134
　　　　　　　3. DNA-DNA Hybridization............................135
　　　　　　　4. Genomic Sequencing...............................136
　　　IV.　Microcosm Tests for Monitoring Techniques.............138
　　　　　A. Microcosm Construction...............................138
　　　　　B. Microcosm Methodology................................139
　　　　　C. Sample Protocol......................................140
　　　　　D. Containment..141
　　　　　E. Points to Consider...................................143
　　　V.　Quality Assurance......................................144
　　　　　A. Introduction...144
　　　　　B. Testing for Sensitivity and Specificity..............144
　　　　　C. Testing for Linkage Between Markers and rDNA.........145
　　　　　D. Summary..146
　　　VI.　Conclusion..147
　　　　　A. Monitoring Techniques................................147
　　　　　B. Scenarios for Protocol Development...................148
　　　　　C. Research Needs.......................................158
　　　VII.　Literature Cited.....................................160

APPENDIX: LIST OF PARTICIPANTS..............................165

Part I

Workshop Summary

WORKSHOP SUMMARY

*Prepared by Gilbert S. Omenn, Workshop Chairman and
Albert H. Teich, Project Director*

A. Introduction and Background

Approximately 60 people, including outside peer reviewers and EPA and AAAS staff, participated in the AAAS/EPA Biotechnology Workshop at Coolfont Conference Center, Berkeley Springs, West Virginia, 29 April - 1 May 1984. As noted by EPA Assistant Administrator for R & D, Bernard D. Goldstein in the opening session, the charge to the workshop was to assess the scientific needs and researchable problems facing the agency as it prepares to deal constructively and responsibly with proposals that might lead to release of genetically-altered organisms to the working or general environment.

The project began with a workshop on 14-16 December 1983 at EPA. Attendees at this meeting included over 50 representatives of EPA program offices (including Pesticide Programs; Toxic Substances; Policy and Resource Management; Drinking Water; Solid Waste; Water; and Air, Noise and Radiation), the Office of Research and Development, and EPA field laboratories. AAAS's role in this first meeting consisted of assisting EPA with the meeting's logistical arrangements and suggesting means of facilitating session deliberations. Most importantly, AAAS involvement in the first workshop facilitated translating the results of that meeting into the basis for the second workshop, as described below.

Based on the deliberations at the first workshop, EPA laboratory and office representatives prepared documents suggesting research approaches which might be necessary to support EPA regulatory efforts in four major areas, which generally paralleled the working groups from the first workshop: health effects, environmental effects, monitoring and quality assurance, and containment and control technologies. Taken as a whole, the four papers constituted elements of a draft biotechnology research agenda.

The second workshop, held at the Coolfont Conference Center in Berkeley Springs, West Virginia, 29 April to 1 May 1984, was convened to subject these papers to an intensive peer review by researchers from outside the agency and by EPA staff, to further define the agency's research plans, needs and capabilities, evaluate them, and modify them to produce an appropriate and feasible research agenda. Names of participants in the Coolfont workshop may be found in the Appendix.

Workshop participants were selected to represent a range of backgrounds, and affiliations. EPA laboratory representatives and headquarters staff brought to the meeting their knowledge of relevant research areas, their understanding of laboratory capabilities, and their own scientific interests and expertise. They represented the interests of those who will actually carry out the research planned at the workshop.

Program office staff represented the perspectives of the clients of users of the research. They brought to the meeting their knowledge of

4 Biotechnology and the Environment

EPA's regulatory mission, their understanding of its needs and priorities, and their views of what needed to be learned in order to carry out the agency's mission and address its priorities.

The outside experts, drawn from universities, government agencies and industrial firms, lent to the deliberations their special expertise regarding scientific aspects of the problems under discussion, their knowledge of other research being conducted in these areas, and their informed opinions regarding the feasibility of proposed efforts.

Participants were assigned to each of the four working groups, and were sent draft documents in advance of the meeting. Each participant brought to the workshop a written review of the paper closest to his or her area of expertise. The workshop began with a plenary session at which summaries of all the draft documents were presented. Brief scenarios of possible EPA involvement in biotechnology, and a summary of the NIH experience were also presented to stimulate discussions.

Most of the meeting was devoted to discussions in four workgroups (panels) corresponding to the four draft papers prepared by EPA scientists:

(1) "Proposed Biotechnology Research Plan for Test Methods Development for Risk Assessment of Novel Microbes Released into Terrestrial and Aquatic Ecosystems" (environmental effects group);

(2) "Proposed Biotechnology Research Plan for Test Methods Development for Risk Assessment of Health Effects Associated with Biotechnology" (health effects group);

(3) "Proposed Biotechnology Research Plan for Monitoring Systems and Quality Assurance" (monitoring and quality assurance group);

(4) "Proposed Biotechnology Research Plan for Environmental Engineering and Technology" (control technologies group).

Plenary sessions were held at the beginning, middle and conclusion of the workshop. The discussions throughout were lively and provocative, going to the root of EPA's role, as well as providing a rigorous review of the proposed data gathering and research projects.

Following the workshop, revised drafts of the papers were prepared and circulated to all participants for comment. Reviews were also solicited from a variety of other individuals who had not attended the workshops. Following this review, it was determined that two of the revised papers did not serve the purpose intended by the workshop organizers, and AAAS and EPA staff agreed to commission new papers to better address the issues. These replacement papers were submitted in draft form and reviewed by EPA and AAAS staff subsequent to the workshop. A preliminary report was prepared by AAAS immediately following the workshop, highlighting the main ideas of the papers and identifying areas for EPA emphasis and attention. Part one of the final report, contains a AAAS summary of the key recommendations and concerns expressed at the workshop. Part two includes revised versions of two of the workshop papers and the two replacement papers: "Monitoring Techniques for Genetically Engineered Microorganisms," by David Glaser and colleagues of Harvard University, and "Biotechnology

Health Risk Assessment Research Plan," by Marvin Rogul of The Rogul Group and John R. Fowle III and David Kleffman of EPA.

B. Key Concerns and Recommendations

Workshop deliberations resulted in identification of a number of areas of concern and recommendations for EPA research activities. No attempt was made to force consensus among workshop participants, but several items were considered high priority for EPA attention. These are enumerated below, and are followed by highlights from the discussions of each working group.

(1) EPA's primary emphasis in biotechnology research should be on potential environmental and health effects of deliberate or accidental release of genetically-altered organisms to the environment. EPA's mandate establishes its lead-agency role in this important area.

(2) Attempts at risk assessment should begin with well-selected specific cases, including, especially, the applications being developed within EPA to control certain pollutants or contaminated sites. A major regulatory need, particularly for OTS, is predictive risk assessment models for products of biotechnology, analogous to the structure-activity relationship models employed for predicting chemically induced effects. It

is, however, premature to attempt development of a general predictive model for assessing the risks of release of genetically altered organisms. Because of the vast number of biological possibilities for biotechnology products (e.g., organisms, vectors, gene sequences, products), it is not possible to predict potential effects without specific knowledge of a number of important parameters. Thus, experience must be gained first on a case by case basis.

(3) Much information pertinent to EPA's biotechnology activities may already be available. The currently available literature should be systematically reviewed, analyzed and used to focus and set priorities for EPA's future efforts. This review will also help foster complementary efforts, and avoid duplication of work performed by other organizations. The scope of the search should include published microbiological and public health information, NIH/RAC sponsored risk assessment efforts, and reports on recent and current molecular biological and ecological studies. The literature should be continuously monitored and collaborative efforts with organizations such as NIH/RAC should be encouraged to keep EPA abreast of developments in the field.

(4) There is a clear need to enhance EPA's molecular biology capabilities and its in-house expertise in biotechnology. A strategy should be developed for gradual, long-term development

of a viable research program capable of attracting and retaining highly competent molecular biologists. Given the Agency's undisputed lead role in environmental assessment, priority should be given to enhancing expertise in this area. Expertise is needed also in such related fields as monitoring of organisms and their products in the environment, fermentation engineering, modern molecular analysis, and health-oriented molecular biology. In general, in order to attract and keep competent molecular biologists, EPA should support some basic research, filling gaps identified in more applied studies.

Close communication and collaborative efforts among EPA scientists with biotechnology-related expertise should be facilitated. Opportunities to discuss research problems and results during weekly seminars and during informal daily contacts foster good science. Spreading modest resources in this area over a number of geographic locations will dilute the talent, perhaps to the point of rendering it ineffective. Consideration should be given to initiating the EPA biotechnology program at a single laboratory site where all ORD office are represented and providing strong management to coordinate activities between offices. It should be noted, however, that EPA's precise personnel needs are inadequately defined at this time.

(5) Outside experts should be identified and called upon to help EPA develop its scientific capabilities in biotechnology, to identify problem areas, to develop and revise research strategies, to peer review research proposals, etc. All of EPA's research proposals should be subjected to rigorous peer review. This workshop set a good precedent.

(6) EPA should actively seek input from industry, public sector interest groups, academia, and other federal agencies on its proposed activities. In turn, EPA should demonstrate to these others that the kinds of approaches and testing any proposed guidelines might include are, indeed, feasible, by leading the way in its own environmental control technology development applications.

C. Environmental Effects

Five research areas were identified which require investigation in order to estimate the actual and potential environmental effects of genetically-altered organisms. The workshop regarded them all as high priority and listed them in the order in which they felt the areas should be addressed:

(1) Methods to identify and enumerate organisms are central to research on environmental effects, as well as monitoring and

health effects. Existing technologies for these functions should be assessed with respect to their applicability to environmental hazards.

(2) <u>Existing methods for measuring survivability and growth parameters</u> should be evaluated to determine where they are effective and whether the kinds of data they yield can be useful in hazard assessment.

(3) <u>The impacts of chemical, physical and biological environmental factors on survival and growth of organisms</u> need to be investigated.

(4) <u>Genetic transfer</u>. Questions in this area, particularly relating to mechanisms, probability, and similarities to or differences from naturally-occurring gene transfer are very important, but they must be recognized as targets for basic investigation, not simply methodology development.

(5) <u>Hazard assessment</u>. Development and validation of methods are needed for studying effects (pathogenicity and infectivity) on non-target organisms. Existing methods for studying disruption and perturbation of environmental processes need to be evaluated as well, particularly with regard to their utility for hazard assessment.

The area of transport was judged to be of significantly lower priority than any of the above topics. The panel believed that existing transport models for various media should be applied first, before any new experimental research is undertaken.

D. Health Effects

The panel felt that because of the large variety of possible organisms and applications, it would not be feasible at this stage to undertake development of a predictive model for health effects. It recommended, instead, a case-by-case approach to estimation of potential health hazards related to release to the environment of genetically-altered organisms. This approach has several aspects (listed in order of priority):

(1) Development of test protocols. Animal test protocols for genetically-altered organisms should be integrated with existing protocols in Subdivision-M of the Pesticide Assessment Guidelines (which already include tests for microbials) and refined. This effort should include, if possible, a scheme for categorizing bacteria according to their degree of pathogenicity. (Subdivision-M also needs to be revised to provide rules for testing viral organisms, particularly those that contain potentially oncogenic genes.)

(2) Data-gathering and information management. Any research effort needs access to a wide range of relevant, up-to-date information. This involves subscription to relevant computer data bases, use of computers to catalog in-house experience, and acquisition of first-rate library resources.

(3) Identification of least pathogenic bacteria. To facilitate both industrial and in-house applications of modified organisms, the agency should identify organisms which are incapable of colonizing man and which may be useful commercially, and should seek to validate that they are indeed safe.

(4) Specific experiments. Among the experiments proposed in the draft paper, those which address possible risks of Bacillus thuringiensis and possible dissemination of baculovirus DNA were considered to be most valuable and feasible. The need to identify genetic markers other than antibiotic resistance for tagging organisms (more generally, the need to conceive and evaluate new genetic markers) was also supported. Such markers would have applications in ecological, health and monitoring efforts.

E. Monitoring and Quality Assurance

Monitoring efforts are central to the conduct of R&D work within EPA. The panel recognized that specific monitoring protocols will need to be developed for individual experiments. In view of this, it felt that the overall emphasis of the monitoring R&D program should be: (a) to evaluate existing tests, (b) to develop new tests for identifying and quantifying genetically-altered organisms in environmental settings, (c) and, based on this experience, to provide guidance on monitoring approaches. Four broad areas of investigation were suggested (in order of priority):

(1) Molecular probes (DNA and RNA probes, immunological methods),
(2) Conventional microbiological methods,
(3) Sampling procedures, and
(4) Quality assurance.

Highest priority was placed on research on the application of molecular methods to monitoring, specifically using DNA fingerprinting methods with microbial populations. This approach possibly could be coupled with the use of DNA and RNA probes to monitor the movement of R-DNA in _in situ_ microbial populations.

The panel also devoted attention to the problem of research needs related to monitoring strategies. While specific strategies should be considered on a case-by-case basis, there is need to compile existing

information on strategies and to develop and test some general
guidelines for planned releases including such factors as site
selection, climatic conditions, sampling arrays, and sampling frequence
inside and outside of the release area. The use of simulants —
analogous organisms whose behavior and properties are well-characterized
— to test alternative strategies was recommended.

F. Control Technologies

The panel took as given that there is a potential hazard in the
release of modified organisms either accidentally or deliberately from
manufacturing processes or field trial applications. It focused on four
problems: (1) assessment of release and exposure; (2) methods for
minimizing release; (3) techniques to prevent worker exposure; and
(4) management of release in situations where it was needed. The
research recommendations fell into two major categories. One set or
recommendations relates to <u>biologically-based manufacturing processes</u>
in which releases can be either accidental or deliberate. These include
(in order of priority):

(1) Studies to assess the potential for release at various points
throughout industrial processes, including evaluations of
individual pieces of equipment;

(2) Assessment of the potential for worker exposure in manufacturing plants;

(3) Evaluation of techniques for containment of specific pieces of process equipment, should there be an accidental release;

(4) Development of monitoring needs and strategies in the context of manufacturing processes;

(5) Assessment of alternative decontamination techniques;

(6) Evaluation of alternative worker protection equipment.

The second set of recommendations concerned deliberate release of modified organisms in <u>field applications</u>. These include (again, in order of priority):

(1) Studies to specify the characteristics of sites which make suitable for various types of field tests;

(2) Evaluation of various alternative approaches to containment materials to be used and interaction of micro-organisms with those materials;

(3) Assessment of monitoring needs and strategies specific to field testing situations;

(4) Evaluation of alternative decontamination methods and materials;

(5) Evaluation of alternative technologies for application of genetically-altered organisms to the environment.

Coordination with other research efforts and the desirability of recommendation to the manufacturing community of guidelines for desirable properties of organisms were also endorsed. Finally, EPA was urged to make its own applications of genetically-altered organisms for control of environmental problems models for the larger research and technical community.

Part II

Papers

RESEARCH PLAN FOR TEST METHODS DEVELOPMENT FOR RISK ASSESSMENT OF NOVEL MICROBES RELEASED INTO TERRESTRIAL AND AQUATIC ECOSYSTEMS

Al Bourquin
Environmental Research Lab., EPA
Gulf Breeze, Florida

Ramon Seidler
Oregon State University
Corvallis, Oregon

I. INTRODUCTION

The Office of Pesticides and Toxic Substances (OPTS), under the authority of the Toxic Substances Control Act (TSCA) and the Federal Insecticide, Fungicide, Rodenticide Act (FIFRA), will regulate parts of the biotechnology industry (1). The major concern leading to this regulatory oversight is uncertainty over human health and environmental effects of organisms released specifically to the environment. This uncertainty comes from the unassessed potential hazard of novel organisms in the environment. In this research plan a series of experimental approaches are presented for developing methods which EPA can cite for measuring risk assessments from the release of novel as well as naturally occurring indigenous microbes. "Novel" microbes include naturally occurring microorganisms placed in environments where they are not native (nonindigenous or exotic), and microorganisms altered or manipulated by humans through techniques of genetic engineering.

The dearth of scientific information on the potential risks from release of novel organisms and the need for a scientific evaluation of risk assessment point to a significant role for EPA's research program in the evolution of the Agency's regulatory scheme for biotechnology.

A. Goal

This document has one major goal: the presentation of a research plan for developing test methods and other information for risk assessment from the effects of novel organisms released to terrestrial and aquatic ecosystems. Test methods need to be experimentally developed so that EPA can cite them in requiring manufacturers to develop relevant data.

The guiding element throughtout this document is to meet OPTS needs for testing methods aimed at identifying hazards and exposures and determining dose-response relationships for novel microbes released to terrestrial and aquatic environments. This focus is justified on the basis of program office needs (2), ORD's research priorities in biotechnology (3), and the expertise of the Office of Environmental Processes and Effects Research (OEPER) in test methods development and exposure assessment. It is clearly recognized that microbes also become dispersed by air currents. However in the initial phases of the risk assessment, containment constraints override concerns for methods development involving air dispersal. Furthermore, it is known that air pollution, as well as chemical and biological warfare defenses have provided mathematical models for describing movement and diffusion of small particles such as microbes. The models of dispersion are generally accepted as an effective replacement of field monitoring for estimating particulate concentrations downwind (65). Therefore, the risk assessment for air dispersal of microbes may not require new methods development.

This document is a research plan and not a research proposal. Many specific research proposals will be developed from this plan and a subsequent workshop to define research to answer some of the more pressing needs of OPTS.

B. Risk Assessment

The basic elements of risk assessment guided the development of this research plan. Risk, defined as a measure of the likelihood and severity of harm (4), is generally assessed through three kinds of investigations: (1) exposure assessment (determining conditions of exposure); (2) hazard identification (attributing adverse effects to the hazard) including dose-response assessment (relating exposure to effects); and (3) risk characterization (estimating overall risk) (4,5).

Exposure assessments make several determinations: (a) the segments of the environment exposed to the agent; (b) the intensity, frequency, and length of exposure; and (c) the concentration and fate of the agent (4,5). Hazard identification, or the process of determining whether exposure to an agent causes an adverse effect, often is a non-systematic investigation (4,5). In fact, a major difficulty, particularly with environmental effects, is in developing tests that will identify the specific impact from the myriad possibilities. Dose-response assessment often requires studies of possible threshold effects and extrapolations from high to low dose, laboratory research to field applications, and few species to many species (4,5). The interdependence among these assessments of hazard (including dose-response) and exposure, is apparent. Finally, risk characterization, the estimated incidence of the adverse effect on a system in a given population, is based on these assessments (5).

C. Two Major Approaches

This research plan presents two major approaches to accomplish the goal of test method development for risk assessment of novel organisms in terrestrial and aquatic ecosystems. For the purposes of this plan, aquatic will include both freshwater and marine (including estuarine) ecosystems, fully understanding that methodology will vary in many instances. Terrestrial will include habitats above (phyllosphere), on the surface, and within the soil ecosystem.

1. Data base development

First, data base development is an important emphasis. The vast and diverse body of ecological and pathological literature on many organisms in many environments contains pertinent information on effects, exposures, and methodologies. We propose to identify, collect, analyze, and evaluate this information to use in test methods development. This objective is consistent with ORD's ranking of data base development as an important priority item in the development of its biotechnology research strategy (3).

2. Research

The second approach for test methods development relies on laboratory research to meet unique needs associated with risk assessment of novel organisms and to complement the information and approaches obtained from data base development. The terrestrial and aquatic microcosms developed by OEPER scientists for the study of xenobiotic degradation are especially well suited for cost effective containment needed for some of the more advanced ("Tier II") studies described in this plan.

D. Short and Long Term Needs

This plan contains both short and long term (one to two year and three-plus years, respectively) research needs and is developed with the following limits and assumptions. The research is planned for microbes, specifically bacteria, fungi and viruses. This emphasis is derived (1) from the greater likelihood that microbes will be exploited for TSCA- and FIFRA-covered biotechnology purposes in the near future; (2) from the likelihood that microbes would be more difficult than macroorganisms to contain in the environment and, therefore, may be a greater hazard, and (3) from the OEPER laboratory's mission for test method development and its current research activities for microbial pest control agents (MPCA). Because of the limitations on research resources and the easy availability of many bacteria already used in gene cloning activities, bacteria have been chosen as the primary model test organisms for early methods development in new aspects of this research plan. An effort to develop the necessary test methods for macroorganisms such as plants would require at the least, resources similar in magnitude to those listed in this document for microorganisms. As the test methods develop and when new agents become candidates for possible release to the environment, it is our full intention to include species other than bacteria in the long term research components. We are well aware of the viruses and fungi which have been released to control various species of pests (57,58). These viruses and fungi will be studied as part of the MPCA risk assessment methods discussed in Section VI. Because the regulatory offices will define the extent of their oversight of organisms released to the environment, no attempt was made in this document to rigorously define novel organisms or genetic engineering. The research plan and test methods can be adapted to whatever groups of organisms become subject to regulation.

E. Relevance to EPA Needs

This plan is presently designed in terms of developing citable methodologies for OPTS regarding terrestrial and aquatic hazards to the environment and will include environmental effects and exposure assessment research. Initial emphasis is on biotic effects within aquatic, soil and above-soil ecosystems. While direct focus on abiotic environmental effects is not a high priority in this plan, consideration is given to the interaction of biotic and abiotic factors.

This research plan was based on the acceptance of the following assumptions: 1) certain information will be available to the regulatory offices concerning the novel microbes. In particular, it is assumed that the regulatory offices will have at least the following information: the identity and source of the microbe, the phenotypic or physiological characteristics of the organism type or at least those characteristics of the unaltered parent, the nature of any genetic manipulation, unique attributes for which it was developed, and intended use including sites, quantity, and manner of dispersal. The eventual use of methods developed under this research plan will be dependent on this information. 2) To conduct a risk assessment of a novel organism OPTS is going to need information on the following (47):

1. ability of the novel organism to survive;
2. ability of the organism to reproduce or persist;
3. ability to be transported to and establish in niches other than those intended;
4. ability to transfer traits to and from other members of its niche;
5. ability to cause adverse environmental effects
    a. pathogenicity, including toxicity and infectivity, to non target species
    b. disruption of environmental processes (e.g. nutrient cycling, nitrogen fixation).

As an initial effort to coordinate the research effort among OEPER's terrestrial and aquatic research groups, a workshop will be held in Washington, D.C. in November 1984. The objective of this workshop will be to assess the state-of-the-art in survival of genetically altered microorganisms and the potential for genetic material transfer in natural environments, including aquatic and terrestrial ecosystems.

As resources in biotechnology become available, scientists competent in the fields of microbial ecology and microbial genetics will be assembled at the workshop to discuss the research areas, to assess accurately the state-of-the-art and to define precisely the approaches to be used by both in-house and potential extramural cooperative researchers. This workshop would center on defined needs and research strategy proposed in the Office of Exploratory Research Biotechnology Workshop and will bring in scientists who have worked in the pertinent scientific disciplines and help to fine-tune research needs. This group would include scientists with whom we could have potential cooperative agreements as well as representatives from other government agencies. By including other agencies, we could increase the potential funding levels and avoid potential duplication of existing research efforts.

II. NOVEL ORGANISMS

The use of novel organisms is growing at a rapid pace. They are being used for a variety of purposes ranging from control of pest populations to the manufacture of pharmaceuticals. Many of these organisms are simply species that are being released into new environments, while others are genetically engineered for specific purposes.

Human health effects research and containment strategies relating to the release of genetically engineered *E. coli* strains have been supported by the National Institutes of Health. The logic developed from that research very early led to the development of specific physical containment and vector requirements for cloning hazardous genes (48). In specific situations, investigators are required to work with debilitated mutants of *E. coli* which cannot colonize the natural environment. However, there is a paucity of information available on the biotic environmental effects associated with genetically engineered organisms of other species. This lack of information is significant and has raised concern (6-8). First, strains designed for purposeful release (including microbiological pest control agents, MPCAs) will, by choice, not be debilitated since they will need to function effectively when released. Second, these microbes will be comprised of species very different, both ecologically and physiologically, from *E. coli*. Furthermore, very little research has been done to assess the hazards resulting from the accidental release of genetically engineered organisms.

Genetic modifications have now been achieved in many classes of organisms ranging from viruses and other microorganisms to higher plants and animals (10). The intent in creating new combinations of genes within a single organism is to provide humanity with new biological tools to benefit health, welfare, and the environment (10-15). Progress in the genetic manipulation of these organisms has exceeded our knowledge of their fate and effect on natural ecosystems (9,14). Before EPA can make decisions concerning release of such organisms, information is needed to establish the potential hazard of engineered microbes in the environment and the consequences of their interactions with other indigenous microbes.

Genetically engineered microbes are being constructed to carry out a plethora of new metabolic activities. Cultures are now able to metabolize new combinations of organic pollutants, to mineralize metals, enhance oil recovery, increase efficiency of nitrogen fixation in leguminous plants, and make animal and human hormones (10,11,12,13,22,23). Very little has been done to assess the ecological fate and effects of such engineered microbes in aquatic, terrestrial and/or agricultural ecosystems (14).

It should be made very clear that human activities have already impacted on the genetic constitution of microbes in the environment. Through the overuse of antibiotics there has been a dramatic increase in the incidence of multiple antibiotic resistance expressed by DNA plasmids which can be transferred into other bacterial species. Various industrial processes including dumping of pollutants and even chlorine disinfection of water selected for strains that are resistant to metals, to antibiotics, and which possess novel metabolic capacities such as the biodegradation of PCB's and chlorinated aromatic pesticides. It is known that transfer of antibiotic resistance, heavy metal resistance, and unique metabolic capacities by cell-to-cell contact occurs in laboratory media, in certain institutions (hospitals), in the animal and human gastrointestinal tract, and in habitats simulating the natural environment (8,13,24,29-31,46). It is most unlikely that the gene pools from novel organisms, which would be released by the billions, will not influence other, new biotic and abiotic ecosystem processes.

III. DEVELOPMENT OF TEST METHODS FOR THE DETECTION, IDENTIFICATION, AND ENUMERATION OF NOVEL ORGANISMS.

A. Statement of Research Problems

The requirement for a technique to identify and enumerate the novel organism following accidental or purposeful release is one of the most fundamental prerequisites of the entire process of risk assessment of novel microbes. Therefore, the development of one or more combinations of specific, convenient, reliable, and sensitive tracer methodologies should be an early if not the first consideration for research needs. Such methods need to be experimentally developed so that EPA can cite them in requiring manufacturers to develop relevant data. The development of these appropriate methodologies is the principal objective of this methods research. EPA will not do actual testing on individual novel microbes proposed for commercialization. A desirable constraint of any detection/enumeration technique is that it be widely applicable and suitable in technical application for detecting any desired novel organism. The test methods that are developed should do little or nothing to alter the mission of the microbe in its intended use of application. The identification and enumeration procedures must discriminate the specific novel strain from other organisms present in the environment and also be capable of discriminating the novel organism from other strains of the same species.

Fortunately, microbiologists have a battery of criteria and techniques for fulfilling these needs. Two broad categories of methods are available. These are: 1) "conventional" selective enrichment methods, and 2) "modern" molecular approaches which rely on the fundamental techniques of genetic engineering. As will become apparent below, there are advantages and disadvantages of both methodologies, and in the final analyses, a combination approach would seem to best satisfy the fundamental criteria for identification and enumeration.

B.   Availability of Data Base

Several modern reference books provide selective media formulations for the recovery of bacteria from their natural ecosystems (49-52). Published references document the usefulness of tagged resistant bacteria as a sensitive tool for monitoring their fate in natural ecosystems.

Applications of the DNA probe for detecting unique DNA sequences in several types of bacteria have demonstrated the reliability and sensitivity of this technique. There is no apparent need for a major effort involving data base development.

C.   Approaches

1.   Conventional techniques

Microbiologists have relied upon the principles of selective enrichment (enhancement) culture and selective/differential media for nearly 100 years for the recovery of various metabolic types of organisms from various ecosystems. With these techniques, the media and other physical/chemical conditions of incubation are adjusted so as to promote the growth of one or more related metabolic types of bacteria. For example, if the novel organism metabolizes cellulose, a medium can be formulated to contain this compound as a sole source of carbon for growth. A broth enrichment medium would be used to detect novel organisms present at counts below the detection limit required for direct plating onto agar media ($\leq$ 100 cells/gm of matter or ml of water). The broth enrichment has the disadvantage of poor direct enumeration but would provide a presence or absence result. In this example, agar media could provide a direct enumeration of cellulose digestors. However, both enrichment media (broth and agar) have the disadvantage of not being specific for the novel strain which was introduced to the environment. Any cellulose degrader in the environment would grow on broth or agar. Therefore it is mandatory to introduce into the novel organisms markers which would increase the specificity of recovery.

The uses of strains which are marked or tagged with a particular resistance or metabolic activity has the advantage of being very specific and allows direct enumeration. There are finite limitations to the kinds of tags available. These tags specifically include resistances to antibiotics, dyes, and heavy metals which can be incorporated into culture media. The novel organism will grow if it is made resistant to these agents. The use of multiple tags will alleviate the problem of a loss of a single resistance marker due to a spontaneous mutation.

Initially, proven laboratory procedures will be used to select spontaneous mutants in chromosomal DNA for the desired resistance. The frequency of spontaneous resistance occurs about $1/10^8$-10^9 cells. At least two such resistance markers should be successively introduced into the novel organism.

The offensiveness of releasing antibiotic resistant mutants can be partially offset by selecting those antibiotics which are not of known importance in treating diseases which the species of novel organism might cause or by relying entirely upon mutants resistant to heavy metals or dyes.

There are certain potential problems with the use of resistant mutants. The effects are unpredictable and must be tested for each novel organism. Potential problems in using organisms with resistance to antibiotics, dyes, or heavy metals include: possible dependence upon the presence of the antibiotic for growth to occur, changes in growth rates, mutability by the selective agent, and possible differences in ability to persist in an ecosystem. Also, certain pesticides (arsenicals, mercurials) may enhance metal resistance in indigenous microbes and mask any such resistance in the novel organism. All these possibilities will have to be experimentally tested to verify that there are no such changes. Specific experiments would involve comparing growth rates in laboratory culture of the parent novel strain and the resistant mutant. Persistence of the two strains will also be compared in simple axenic ecosystems as well as in simple ecosystems containing the natural microbial flora.

2. **Molecular techniques**

There are several molecular techniques which are useful for the identification of organisms. It should be noted, however, that the molecular techniques do not have the ability to enhance selective growth nor do they provide a selective, differential milieu which can be provided by conventional techniques. The molecular techniques will not be useful if the novel organism has declined in numbers to the point where it cannot be directly detected on a suitable agar plating medium (limits are < 100 cells/gm matter or ml of water).

A DNA gene probe, specific for the DNA base sequences present in the novel organism, provides a powerful, specific, and reliable detection and enumeration tool (54,55). The DNA probe must reflect DNA sequences which are unique to the novel organism. This can be accomplished by either preparing the probe to reflect a novel plasmid or a smaller portion of a custom DNA sequence incorporated into a plasmid in the novel organism. The probe can be radiolabeled and hybridized to colonies of the novel organism growing on the surface of a nitrocellulose filter in contact with an appropriate agar medium. By radioautography one can locate and count colonies of the novel organisms present on the original filter surface. This technique is relatively expensive and not as convenient as a conventional direct selective plating medium used in conjunction with tagged strains.

A second molecular technique involves the introduction of a plasmid containing DNA sequences which will respond upon command to produce (induce) a visible chromogenic substance by the novel organism. Examples would include the induction of an enzyme which causes a color change in an ingredient in the culture medium or induces the production of a pigment such as the red-colored prodigiosin compound. The experimental details of such a plasmid and its induction mechanism have recently been described (56). The induction of the chromogenic substance triggered only upon application of a chemical or a temperature change would provide a means of distinguishing the novel organism from any resident microbes which may also possess the same metabolic capacity.

Once the novel organism is isolated and obtained in pure culture, its plasmid DNA could be isolated and "fingerprinted" with specific types of restriction enzymes (16,53). This will verify the identity of the unique DNA fragments as those present in the novel organism prior to its release. This fingerprinting procedure would serve as a verification check that the selective culturing conditions were indeed specific for recovering the unique novel organism of concern.

D. Short Term Products

Early efforts should be directed at compiling selective and differential media formulations from published modern literature appropriate to novel organisms which are likely to be released. There are several recent publications available to accomplish this task. Another early effort should involve inducing resistance markers in novel organisms and examining them for growth rate effects and survival capabilities in "simple" ecosystems. Persistence and growth rate studies of the resistant and parent novel organisms should be compared. Appropriate quality assurance parameters should be established for all identification and enumeration assays using EPA recommended guidelines. Initial studies should be conducted in conjunction with part IV, fate and transport, to examine novel organism recovery and detection efficacies from suitable model ecosystems.

E. Long Term Products

Test the efficacy of detection procedures using molecular techniques, with initial efforts devoted to DNA probes which are already available for specific organisms. Test applicability of DNA probe for recovery and detection of novel organisms specifically from terrestrial and aquatic ecosystems. Evaluate how conventional and molecular techniques can be best utilized and how procedures meet expectations of the quality assurance program guidelines (validation of test methods). Prepare technical papers on performance and validation of test methods.

32 Biotechnology and the Environment

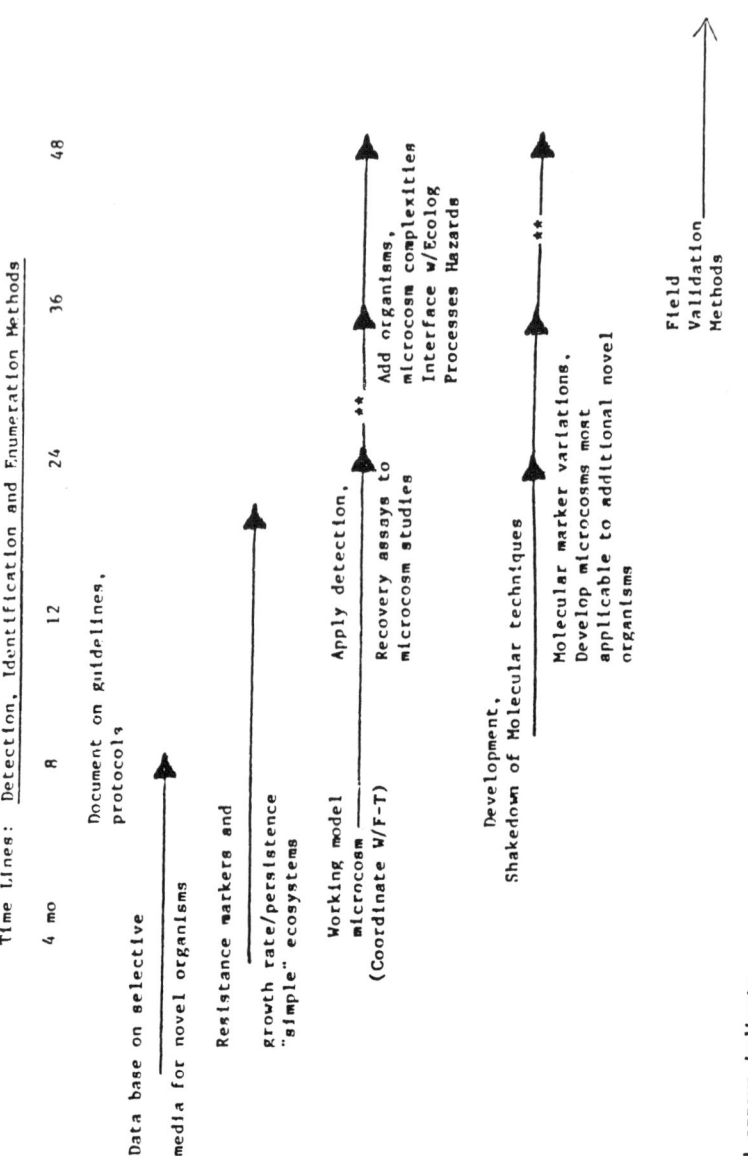

IV. DEVELOPMENT OF TEST METHODS FOR ASSESSING FATE OF NOVEL ORGANISMS

A. Statement of Research Problems

Test procedures must be designed for determining whether novel organisms persist or multiply to high levels and therefore increase the risk of escaping from the niche for which their use was intended. Experimental aquatic and terrestrial ecosystems for studying persistence and fate will also provide information on biological systems and ecological processes which become exposed to or affected by the novel microbe. Such information will be valuable in selecting relevant methods to deal with infectivity and pathogenicity and ecosystem perturbation models in long-term research evaluating exposure-risk phenomena appropriate for novel microbes (Section VI).

Furthermore, persistence (monitoring population changes) and ecological fate (niche and physical location) testing methodologies will be useful and applicable in developing protocols for validating genetic expression in complex ecosystems and for delineating biotic and abiotic factors which contribute to survival or nonsurvival of the organism or its genetic traits.

The choice of the model ecosystem will be predicated on practical criteria which can fulfill the following set of conditions. The model ecosystem must be flexible in design and capabilities to accommodate the intended use of a spectrum of novel organisms. It must be housed in such a way as to prevent accidental release of novel organisms during and after the test period. This requirement imposes size restrictions on the type of ecosystem and the number of experimental units employed. For example, since plants will be a biotic component in terrestrial ecosystems, regulation of photoperiod would be mandatory. The test units should accommodate appropriate physical and biotic trophic components i.e., a terrestrial unit should be

capable of housing plants, invertebrates (chewing insects, earthworms, etc).
Similar concerns for the design of aquatic units are also apparent.
Furthermore, as part of the proposed experimental protocols, it is envisioned
that various procedures will be used to test, on a worst case basis, changes
which can prolong the survival or encourage growth of the novel organism.
Such perturbations might involve the addition of decaying plant and animal
matter in both the terrestrial and aquatic systems, as well as variations in
plant and invertebrate stocking densities. It is clear that a variety of
simple (flasks, mason jars, aquaria) and complex (microcosms) containment
systems are available for fulfilling the necessary test methods criteria.

B. Availability of Data Base

A vast amount of data base information is available on containment
systems used for assessing fate of chemicals. The EPA has been involved in
most of the significant contributions in this area (25-27) and OEPER
laboratories have developed and are using such systems for both aquatic and
terrestrial research. These containment ecosystems are adaptable to the
present needs. A review of the vast amount of literature should be made to
derive the most appropriate experimental designs for application to the
present tasks.

Studies conducted over some 60 years are available dealing with the
persistence of certain microbes in natural ecosystems. Most of this
literature deals with organisms associated with animals and plants (pathogens
and pollution indicator bacteria). Other, more contemporary information needs
to be summarized on persistence, cell densities, and fate dealing more
specifically with genera of organisms likely to be released including species
which are associated with the plant phyllospheres and root rhizosphere
(_Pseudomonas_, _Xanthomonas_, _Erwinia_, _Klebsiella_, etc.) and those likely to be

found in fresh and marine water systems (Pseudomonas, Vibrio, Nitrosomonas, Alcaligenes, Actineobacter, Aeromonas, etc.).

C. Approaches

1. The microcosm approach

Initially a series of method developments are envisioned for monitoring survival of novel organisms in simple containment systems such as flasks, mason jars, membrane filter chambers, etc. In later, more complex risk assessments, and for purposes of containment, scientific control, habitat simulation and convenience of experimental manipulations, tests should be conducted in microcosms. The microcosm envisioned for terrestrial research (43) uses a soil/plant/water ecosystem with water runoff collected in an aquarium. Thus the fate of soil, plant and invertebrate-associated microbes which are transported through the soil into the aquatic ecosystem can also be readily monitored. The transport and fate of the novel organism in fish can also be investigated. The terrestrial/plant microcosms will be maintained to simulate growing conditions and soil temperatures and moisture can be readily controlled. As resources permit, seasonal conditions will be selected to allow plant senescence and the accumulation of decaying animal and plant matter. Two different kinds of microcosms are appropriate for studying the fate of these organisms in aquatic environments. These include intact sediment cores and flow through chemostats. These devices will contain both micro and macroorganisms appropriate for studying fate of novel organisms in aquatic ecosystems.

2. Rationale for selecting ecosystems
    a. Terrestrial research

   Attention will be focused on five separate kinds of experimental approaches for assessing fate. All experimental approaches will be conducted in a terrestrial model microcosm which the EPA has developed. The five ecosystems to be investigated in experimental microcosms are:

1. Rhizobium/legume soil ecosystem.
2. Root rhizosphere of easily manipulated and cultivated plants (radishes and potatoes).
3. Soil/plant ecosystem involving fate and effect of engineered microorganisms capable of metabolizing pesticides.
4. Vegetables undergoing microbial decay simulating remains left in fields following commercial harvest.
5. Plant leaf surfaces.

Since terrestrial ecosystems will be used, novel organism fate and transport will be monitored not only through and on plants and soil, but through other biotic components such as insects and earthworms and mammals, (surfaces and intestines) as well as decaying plant and animal matter.

The research strategy has been built around a selection of those unique natural terrestrial habitats which support high natural cell densities in nature. For example, inside an average legume nodule approximately 10^7-10^9 viable cells are found (32,33). Plant root rhizospheres contain approximately 10^6 or more bacteria per cm length of root surface (34,35). The spray application onto soil and plants of herbicide-metabolizing bacteria (11,12) will probably not initially achieve high cell densities unless there is significant regrowth of the organism. However, it is envisioned that application of pesticides will achieve two unique conditions. First, many of the indigenous soil bacteria will be killed by the pesticide making both nutrients and niches available for the resistant bioengineered or plasmid containing microbe. Second, the vegetation will undergo natural decay and the remaining indigenous soil microbes will greatly increase in numbers. This will allow growth of bacteria of many genera including Pseudomonas, Erwinia, Klebsiella, Enterobacter, Alcaligenes, Pectobacterium, and others. Representatives of these genera are involved in bioengineering efforts and can easily become candidates for release to the environment. Finally, it is common knowledge that the commercial harvest of vegetables leaves behind hundreds of pounds of plant material per acre. As this material decays, there will be a great increase in selected portions of the terrestrial microflora, especially those plant pathogenic soft rot bacteria of the genera Erwinia, Pectobacterium, and Pseudomonas. Representatives of these genera and other Enterobacteriaceae are well known for their promiscuity involving plasmid transfer (10,14,16,31,34) and could easily come into contact with the novel organism.

b.   Aquatic research

Studies with aquatic systems may utilize "Eco-cores" which consist of sediment cores taken from the bottom of lakes, streams and estuaries (59-61). These cores are currently used to determine degradation rates of xenobiotics by the organisms present in them. These microorganisms, as well as mixed culture chemostats (12), can be employed in studies designed for investigating consequences of novel organism release into aquatic ecosystems. Also, membrane chambers can be used to develop methods for studying persistence and die-off kinetics of novel organisms in self-contained aquatic ecosystems containing various levels of nutrients.

Much consideration has also been used in selecting those unique aquatic ecosystems which support naturally high densities of microbial populations. Ecosystems with high microbial densities are those most likely to foster occurrences of gene flow between species components. There are a number of ecosystems which support high microbial densities that can be experimentally evaluated in the eco-core microcosm. These include certain aquatic sediments, Spartina salt-marsh sediments, and various liquid/surface interfaces. Certainly the evidence of genetic transfer in nature between organisms appears to infer that cell-to-cell movement of genes depends more on ecological intimacy than on evolutionary relatedness and the proposed aquatic microcosms will be designed to provide the opportunity for detecting gene transfer events.

There are additional justifications for selecting these ecosystems for the study of fate and transport. In many cases it has already been demonstrated that these environments contain microbes important in agricultural processes (11,13,32-42). Microbes from these environments have already been subjected to much research, often involving gene manipulation and bioengineering. It is likely that bacteria from these environments will someday be purposefully released to solve applied biological problems associated with agriculture and pollution abatement (62). In some cases, gene transfer between plasmid containing strains of Escherichi coli and other bacteria obtained from these environments has already been demonstrated

(28). Finally, many of these ecosystems contain bacteria which can become pathogens (or opportunistic pathogens) of humans, animals, and plants.

D. Short Term Products

The first short term milestone will involve the preparation of a document which deals with designs of microcosms appropriate for the current task. Similar documents already exist and this task should not require much effort. In addition a document will be developed to summarize available literature on survival of novel species which are prime candidates for release to aquatic/terrestrial/plant environments. A suitable containment laboratory will be designed and equipped with the experimental microcosms. Investigators will establish which novel organisms to use in initial evaluation of protocols for fate and transport; they will establish appropriate choices of higher plants and invertebrates and the appropriate range for stocking densities. Physical and environmental factors which contribute to transport and survival of novel organisms will be measured in two or more of the proposed ecosystems. However, initial studies will rely heavily on more simple and inexpensive experimental ecosystems (soil-jar, aquaria, etc.) to evaluate general physical/chemical/nutritional factors influencing survival and fate of novel organisms. Conditions will be identified which influence organism persistence or growth in simple systems and used for specific experimental design considerations in developing the more complex microcosms. The scientific importance and relevance of the microcosm approach for studying fate and transport of novel organisms must be established. This will be accomplished by publishing technical papers dealing with survival of novel species which are prime candidates for release; by publishing reports on the description and use of the microcosm with preliminary scientific data establishing design specifications, containment, and initial data. OEPER should sponsor a workshop program at a national scientific meeting on, "Ecosystem approaches for studying novel organism fate and transport".

E. <u>Long Term Products</u>

The microcosm experimental approaches will be expanded to include other ecosystems and also reflect the needs of new developments of novel organism applications. Investigators will establish parameters necessary for fulfilling quality assurance and data validation and publish additional technical reports. OEPER will organize a workshop with environmental and microbial ecologists to critique experimental protocols and discuss state of the art alternatives for simulating natural ecosystem processes for fate and transport studies in terrestrial and aquatic ecosystems. The proceedings of this workshop will be published in the EPA Technology Series.

Test Methods Development for Risk Assessment of Novel Microbes

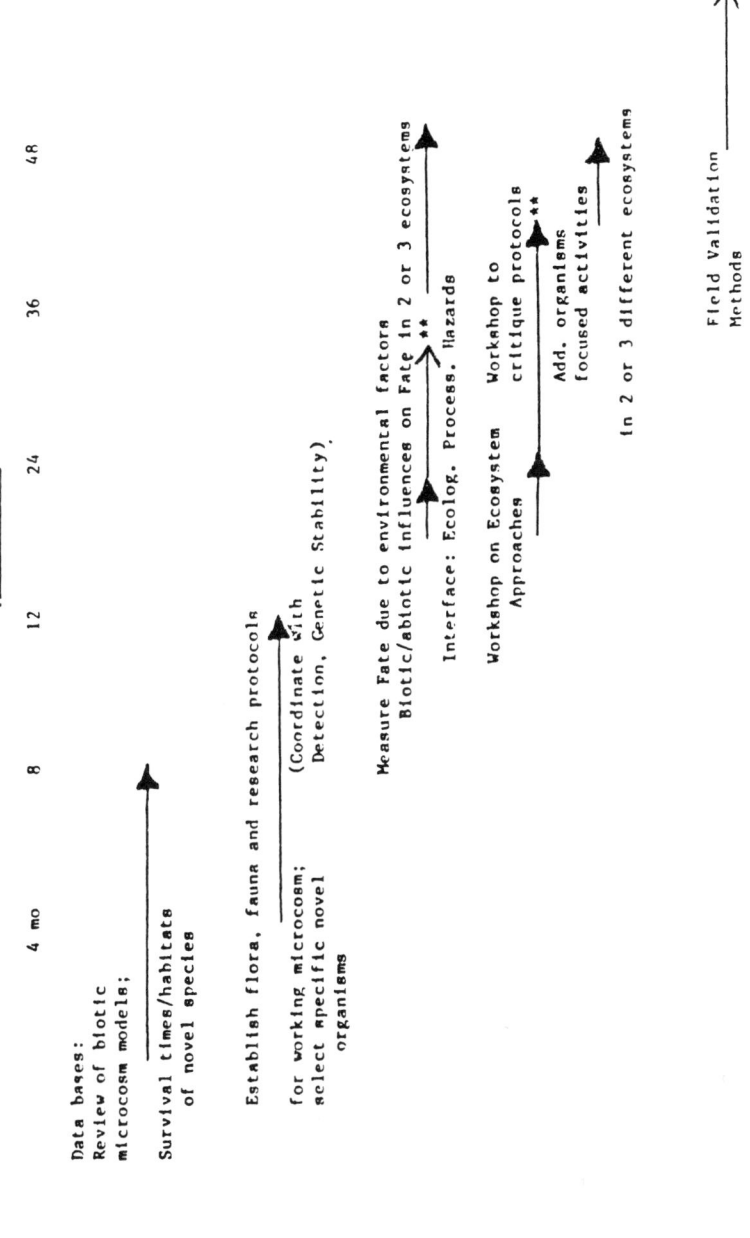

V. DEVELOPMENT OF TEST METHODS FOR ASSESSING GENETIC STABILITY OF NOVEL ORGANISMS

A. Statement of Research Problems

Test procedures must be developed to quantitatively measure transfer frequencies of genetic material from and into novel organisms which enter the environment. Suitable systems must be established so that genetic traits of novel organisms can be monitored for stability. The transfer of genes into new species makes it exceedingly difficult to monitor fate and effects since we must now monitor gene pools and new organisms. If genetic traits are transferred into microbes which have ecosystem functions different from the novel organism, new abiotic and biotic components of that system may be adversely impacted.

It should be mentioned at the outset that assessment of genetic stability will be largely influenced by results obtained in companion studies discussed in this document. The level of concern or risk involving genetic stability or lack thereof will, for example, increase or decrease dependent upon results of the persistence, fate and transport, and hazard assessment data. If, for example, it is found that the population of specific novel organisms declines rapidly (hours to days) and never regrows following entry into the environment, it is more unlikely that concern will arise from its genetic capabilities. If, on the other hand, an organism colonizes an ecosystem and achieves sufficient cell densities to affect on other biotic processes, an increased level of concern will arise over the fate of its novel genetic information and the possible impacts or hazards that might result from it.

A major research challenge is formulating an all inclusive approach to assess the ways genetic stability is to be measured. The approach must consider the myriad of possible ways an organism can be bioengineered or otherwise genetically altered, the kinds of species involved, the ways in which gene transfer can occur, and that commercial products may contain not just one but rather a consortium of genetically and taxonomically undefined organisms.

It is also recognized that physiochemical characteristics of different types of soil, water, and other environs significantly affect the fate and genetic stability of novel organisms. Consequently, considerable attention should be focused on studying a broad spectrum of simple microcosms so that as many physicochemical conditions as possible can be examined.

The research will build upon existing information on microbial genetics, DNA plasmid biology, and a synthesis of fundamental principles of microbial ecology. A multitier approach is proposed, proceeding from simple, rapid laboratory experiments to the habitat simulated microcosm analyses.

B. Availability of Data Base

Data base needs indicated earlier for persistence, fate, and specific environmental components impacted by various species will be of use in the present study of genetic stability. In addition, information must be gathered on the published information concerning gene transfer capabilities of major bacterial groups. This should include information on whether chromosòmal DNA can be transferred and a compilation of functions and types of plasmids which have been found in the representative organisms. Similarly, this review should include information on other types of DNA transfer mechanisms such as transformation and phage mediated transduction. A compilation of gene transfer frequencies, the kinds of gene functions which are transferred and the species which might serve as recipients of genetic information from novel organisms will be made. Much of the previous research on these subjects is limited to bacteria which are animal or plant pathogens, but extensive information is available for bacteria of the genera Escherichia, Pseudomonas, Bacillus, Acinetobacter, and Rhizobium. Assembling this data base will be no small task, and a number of scientific laboratories should become involved. However, the work should be distributed to the reviewer with expertise on the particular genus or group of bacteria.

C. Approaches

1. Naked plasmid DNA

In the first phase of genetic stability studies, the fate and effect of naked DNA plasmids released to the microcosm will be investigated for both maintenance of their original physical characteristics (molecular weight, covalently-closed circular nature, nicks, and enzyme restriction patterns) and for biological activities (ability of released plasmid to transform competent suspensions of E. coli). The ability of resident strains in the ecosystem to take up selected plasmid DNA molecules will be investigated.

There is some debate as to whether experiments on the fate and effect of naked DNA is important. Some scientists suggest that naked DNA will be quickly inactivated by DNA hydrolyzing enzymes present in the natural ecosystem. The experimental approaches which follow are intended to carefully verify this possibility. In preliminary control experiments plasmid DNA will be added to various simple ecosystems and its stability will be evaluated by periodic extraction and testing for physical and biological activities. If the DNA is not stable, additional experiments would not be conducted.

If the DNA is found to maintain physical integrity, the following approaches will be used. The fate of plasmids of different molecular weights and incompatability groups will be investigated. The reviewed literature (14) as well as two research publications (44,45) have already demonstrated the feasibility of such studies and have commented on the importance of establishing the fate of naked DNA in microbial ecosystems. The relevance of conducting such experiments lies in learning the biological fate of plasmid DNA released from lysed bacterial cells. It is also important to know the fate and effect resulting from the release of plasmid DNA into specific niches which could be used to genetically transform indigenous microbes to carry out agriculturally relevant activities (increased legume host range, antibiosis in root rhizospheres, etc.) or those organisms responsible for various degradation activities in water and soil systems.

The fate of naked plasmid DNA will be investigated in:

    A. Soil.

    B. Plant associated habitats (plant root rhizospheres, plant phyllosphere, decaying plants, legume nodule environments).

    C. Fresh and estuarine water and associated habitats.

The fate of naked DNA will be followed in indigenous bacteria (see page 22) including selected species of plant pathogens (Pseudomonas syringae, Agrobacterium).

2. <u>Stability of plasmid DNA in novel organisms</u>

In the second phase of studies the stability of the genetic material carried in novel organisms will be investigated with regard to its transmissibility to indigenous bacteria.

In these initial Tier I tests, the novel organism will be grown to high cell densities in laboratory media and combined in separate tubes, with the bacteria it would be likely to contact in nature. Using routine selective media, an examination will be made for recombinant organisms, i.e., those which have received marker genes from the novel organism. Calculations of transfer frequencies will be documented (recombinants per donor cell). In appropriate cases, transfer of chromosomal DNA will also be examined.

In Tier II studies, organisms exhibiting gene transfer in Tier I will be placed into terrestrial and aquatic test systems as described in Section III, fate and transport. A similar study will be conducted here, with the purpose of examining for recombinants among the microbes present in the ecosystem. Gene transfer frequencies will be documented.

The fate in various novel bacterial species of conjugative and nonconjugative plasmids of different incompatibility groups will also be investigated in both Tier I and II studies. These studies will employ genetically characterized broad host range plasmids as well as "safe" nonmobilizable plasmids. Bacterial species chosen as prospective donors (potential novel organism candidates) are appropriate to the different ecosystems under study:

- A. The soil ecosystem (ex. Pseudomonas, Alcaligenes, Klebsiella, Enterobacter, Rhizobium).
- B. The aquatic ecosystem(s) (ex. Pseudomonas, Escherichia, Klebsiella
- C. The plant-associated habitats (ex. Pseudomonas, Alcaligenes, Agrobacterium, Citrobacter, Enterobacter, Erwinia, Pectobacterium, Klebsiella).
- D. The invertebrate and mammalian ecosystems (genera as above) plus Escherichia coli.

Potential recipients of genetic material will consist of genera present in the ecosystem intended to receive the novel organism.

All mating combinations investigated in microcosms will be initially tested in the laboratory for feasibility and frequency of occurrence. Screening systems will also be tested for the detection of appropriate biological markers in recombinant organisms. Plasmid containing strains will serve as donors while recipients, in this initial phase, will be those naturally occurring strains isolated from the various ecosystem models in which the novel organism will be released. In addition, well characterized strains of E. coli will serve as control recipients.

In the microcosm analyses the occurrence of triparental matings will be measured in addition to the direct plasmid transfer between two participating cells. Triparental matings will provide an index of the incidence of helper plasmids in the terrestrial microcosm flora and will define the capabilities of natural helper plasmids to mobilize different classes of plasmids present in the novel organisms added to the ecosystems. This aspect of the study is especially important in assessing the fate and effect of novel microbes and has not been investigated before. Plasmid containing donor cells added to ecosystems will be constructed, as far as applicable, to contain various combinations of a conjugative plasmid or a nonconjugative plasmid as well as those which are or are not mobilizable, as well as strains which contain pairs of different plasmid types.

Certain genetic elements are designated by the term, transposons (16,17,20). Transposons or "hopping genes" have the ability to leave a plasmid and enter the chromosomal DNA of the organism which contains it or can be transferred into another bacterium if carried on a mobilizable plasmid. Transposons are known to occur among microbes in nature. Therefore, consideration should be given to the importance of transposons in transferring genes from a novel organism plasmid to its chromosome or vice-versa. The subsequent transfer of these genes to other indigenous organisms could be enhanced by the occurrence of transposons and helper plasmids in the natural microflora. Research elements should be investigated for developing citable methods for evaluating the roles of both transposons and helper plasmids in assessing the genetic stability of released novel organisms.

In one group of the terrestrial in vivo mating trials, Rhizobium strains will be used which nodulate subclover or soybeans. Inoculations of seeds will be conducted using two Rhizobium species or strains which have the potential to occupy the same legume nodule (double infection; 32,33). In this manner the possible in vivo transmission of plasmids within individual legume nodules can be measured.

In addition, specific strains of Alcaligenes and Pseudomonas will be used as potential donors of plasmids which code for the metabolism of 2,4-D biodegradation (13,46). These plasmids have already been shown to have a broad host range and are mobilizable to a variety of bacterial species, at least under laboratory conditions (13,40).

Screening and detection of recombinants will be made possible through the use of genetically tagged donor strains. Recovery of recombinants will be achieved by plating out specimens on selective media containing, for example, antibiotics appropriate to the plasmid resistance markers, and a carbon source not utilized by the donor strains. Screening and detection of recombinants will be facilitated through the use of appropriate markers induced by transposon mutagenesis (17,20).

Presumptive in vivo plasmid transfer will always be verified. This will involve demonstrating the taxonomic or phenotypic distinctiveness

between donor and recombinant as well as actual physical demonstration of
the appropriate plasmid in recombinants. The latter will involve
characterization of plasmid molecular weight and enzyme restriction
patterns. Recombinants will also be examined to determine if they can in
turn serve as donor cells to transmit the plasmid which they have
received.

3. Sources of cultures and plasmids

Bacterial cultures will be obtained from several sources. These
include specific cultures, already bioengineered, requested from
scientists who have described such strains in the scientific literature.
Cultures will also be obtained directly from the environment (plant root
rhizosphere, decaying vegetation, sediment, etc.) and from international
culture collections. Cultures isolated from the environment will be
characterized taxonomically to at least the genus level. These strains
will serve as experimental donors and recipients of plasmids representing
various types of conjugative and nonconjugative groups which may be
employed in the construction of various kinds of novel organisms.

Plasmids are widely available from international culture collections,
biological supply companies, or other scientists, and can also be
constructed in the laboratory using basic bioengineering techniques.

D. Short Term Products

The first short term products will be the technical documents which
compile known genera which have gene transfer capabilities. It is envisioned
that information will be compiled by genus to include not only chromosome,
plasmid, naked DNA, and bacterial virus gene transfer mechanisms, but also the
precise methods under which DNA exchange was established (growth phase, cell
densities, temperature, etc.). Tabulated information on the habitats where

the organism resides and whether literature is published on natural ecosystems where DNA transfer has been documented will also be obtained.

Initial research will be conducted in soil and leaf phyllosphere habitats to investigate the stability of naked plasmid DNA.

Tier I test procedures for conjugation and plasmid DNA transfer will be initiated. Initial protocols will use common bacteria which are likely candidates for release (Klebsiella, Xanthomonas, Pseudomonas) and which are known to exhibit gene transfer capabilities. Working protocols will be established.

Quality assurance protocols will be developed in parallel with research developments.

E. Long Term Products

Technical reports will be published on methods developed including fate, effects, and stability of naked plasmid DNA.

Tier I DNA transfer experiments will be continued using various types of DNA plasmids. The host range of plasmid transfers (donor-recipient combinations) will be based on intended uses of novel organisms and organisms in habitats impacted. Experimental niches will be established and tested for triparental matings in terrestrial microcosms.

EPA will sponsor a workshop at a national professional meeting on, "Prokaryotic gene transfer in natural ecosystems: relevance to release of novel organisms." Appropriate technical reports will be published and all appropriate methodologies will be validated.

Test Methods Development for Risk Assessment of Novel Microbes 51

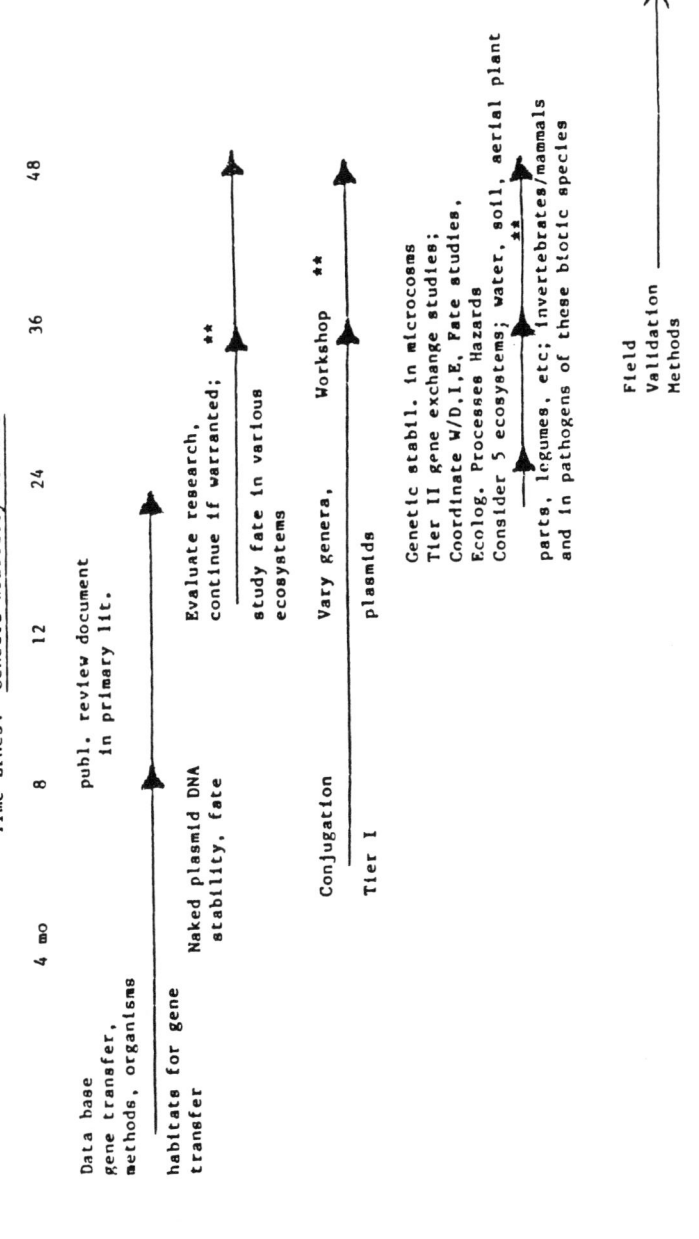

VI. DEVELOPMENT OF TEST METHODS FOR ASSESSING HAZARDS OF RELEASED NOVEL ORGANISMS

A. Statement of Research Problems

The regulatory goal of permitting the release of only low-risk novel organisms requires the development of test methods that will distinguish between hazardous and non-hazardous organisms and test methods which allow a determination of the nature of the hazards if they exist. This portion of the research plan is organized around two major kinds of potential environmental hazards from novel organisms released to the environment: 1) pathogenicity, including toxicity and infectivity for various non-target organisms and 2) other effects on environmental processes.

This approach to hazard identification is based on the assumption that a variety of test methods will be developed applicable to a variety of intended-use ecosystems. The selection of methods to which to subject a given organism would be based on the knowledge of the organism's identity, source, unique attributes for which the organism is released, genetic manipulations, and intended use (site, concentration, application method). Portions of the hazard identification methodology will be dependent upon exposure assessment for determining populations exposed to the agent. The latter information will be available from experimental data obtained from Section IV on fate and transport.

There are two categories of research problems which reflect the two kinds of environmental hazards. First, there is a need to develop test methods for pathogenicity, including toxicity and infectivity of novel organisms released in terrestrial and aquatic ecosystems; second is the need for test methods for determining the nature, magnitude, and consequences, if any, of novel microbes on natural ecosystem processes other than pathogenicity and gene stability.

It is anticipated that data requirements in subdivision M (63) would be useful for screening novel organisms (MPCAs and non-MPCAs) for pathogenicity, toxicity, etc. Simple and complex containment studies would be the approach for assessing hazards to ecological processes.

B. <u>Availability of Data Base</u>

The Office of Pesticide Programs (OPP) has developed guidelines for testing hazards to non-target organisms by microbial pesticides. These guidelines provide detailed methods for testing pathogenicity, including toxicity, and infectivity, of microbial pesticides to a variety of organisms from terrestrial habitats (63). In addition, a vast amount of information available in the plant, animal, and microbial pathology literature would provide methods and data concerning other (non-pesticidal) microbes. Guidelines for similar testing strategies involving engineered MPCAs are predominantly the same as for non-engineered MPCAs (64).

On the other hand, OPP guidelines on ecosystem processes and expression data requirements for microbial agents are relatively undeveloped and, therefore, will be of very little value in this test methods development. However, there exists a vast literature in plant, animal, and microbial ecology and environmental microbiology which contains experimental approaches, and test methodologies pertinent to ecosystem processes.

There is no similar data base for organisms that may be used in industrial processes that are covered under the Office of Toxic Substances Programs.

C. <u>Approaches</u>

Short-term research is directed at a) data base development and b) an evaluation of initial experimental data obtained from fate, transport and survival studies developed from Section IV.

In the first approach, a search and evaluation will be conducted of literature for information and methods for testing pathogenicity, toxicity, and infectivity, of novel microbes to various organisms in target ecosystems and in ecosystems to which they are transported. A small workshop with program office personnel and outside experts will be held to provide guidance on naming likely novel microbes and identifying model(s) used for developing test methods for measuring hazard(s) to target and non-target ecosystems. The workshop should also assess applicability of OPP non-target hazard guidelines to TSCA-covered novel organisms. Recent assessments of OPP guidelines should be helpful in this applicability study. Goals of the workshop would also be to identify the applicability of and/or changes in guideline tests for other novel organisms.

The second approach, will be to develop methods of hazard assessment based on microcosm experimental data obtained from fate and transport and genetic stability studies. This data will provide insights for formulating models of risk assessment. Possible hazard risk assessment models currently considered as candidates include: a) involvement of aquatic invertebrates and insects as vectors which transport novel agents to and from target and non-target plant species including plant pathogenic viruses, nuclear-polyhedrosis virus, bacteria, and fungi; b) application of a novel organism to a toxic waste applied to soil or water and the toxic waste or its degradation product is mutagenic; c) use of a novel organism with increased capacity to mineralize matter that results in a soil pH change affecting plant growth; or reaches significant numbers in terrestrial or aquatic systems and influences a natural biogeochemical process (C, N, S cycles); d) niche displacement of microbes associated with plants, insects, etc. by aggressive colonizing novel bacteria. These approaches to hazard assessment applicable for target and

non-target organisms will include the spectrum of biological species
(microbes, plants, invertebrates, birds, mammals).

Longer term research (laboratory, microcosm) in this area would be of
three kinds: (1) continue to revise and test the OPP hazard guidelines for
microbiological pest control agents (MPCAs); (2) fill gaps in ecological
process test methods which are apparent after completion of short-term data
base development and evaluation; (3) validate test methods.

D. Short Term Products

OEPER will conduct the workshop and prepare a report on the novel
organisms most likely to be used and their most likely route of release and
the ecosystems impacted. A document will be prepared on data base development
and evaluation which contains methods and other pertinent information from
literature searches.

OEPER will prepare a separate report evaluating applicability of OPP
guidelines to TSCA covered novel organisms containing recommendations and
modified guidelines with indications of gaps in information. A search and
evaluation will be conducted of the ecological literature for information and
methods for testing ecosystem expressions and processes. It will also be
important to delineate discrete ecosystem types and discrete testable end-
point effects in the various ecosystem types. Guidance should be available
from the program office.

Five to 10 ecosystem processes will be identified that are likely to be a
target of novel organism risk assessment. Then it should be possible to
delineate possible end-point effects within these ecosystems. Measurable endpoints should exist, for example, in the systems of xenobiotic degradation,
nitrogen, carbon, and mineral cycling; in mutualistic relationships such as
Rhizobium-legume and mycorhizzal associations; in competition in the

rhizosphere and phyllosphere of plants (changes in resident populations, niche displacement, effects on trophic levels).

E. <u>Long Term Products</u>

Investigators will publish technical papers showing methods development dealing with examples of hazards (pathogenicity/toxicity, ecological process changes) and their measurement. Technical papers will also detail validation of test methods. OEPER will prepare a set of suggested guidelines for OTS and OPP.

A significant goal is to develop research systems in which to test for effects on various endpoints as delineated in short-term research. We would anticipate that the most likely approach will be with the use of microcosms. However, the test system will be developed to gain information about specific endpoints and not ecosystems in general. As methods proceed and develop, there will be appropriate validation of all tests.

Test Methods Development for Risk Assessment of Novel Microbes 57

Summary Time Lines, All Test Methods

		OTS PROTOCOLS, SUPPORT DOCUMENTS:
Initial Workshop November 1984		
Data base documents All sections		6-10 mo
Established working model microcosms		12 mo
Detection, Identification, Enumeration		
	Conventional Techniques	36 mo
	Molecular Techniques	40 mo
Fate		
	Microcosms	24 mo
	Biotic/abiotic factors on Fate and Transport	36 mo (Draft)
	Post workshop updated document	42 mo
Genetic Stability		
	Fate/Effect naked DNA	34 mo
	Fate/Effect plasmid DNA	
	Tier I	30 mo
	Tier II	36-40 mo
Assessing Hazards	MPCA	
	Insects, Avian	28-30 mo
	Mammals, Plants	40-48 mo
	Ecological Processes	36-48 mo

VII. SUMMARY

The goal of this document is to present a research plan for developing test methods for risk assessment involving the accidental or purposeful release of novel organisms to the terrestrial or aquatic environment. The document is not a research proposal but many detailed research proposals will be developed from the plan.

The plan is developed on the basis of program needs, OEPER's research priorities in biotechnology, and the expertise in test methods development in our research group at various laboratories.

There are two major approaches to the plan: data base development and research. The scope involves both short (1-2 years) and long term needs. The methods development addressed regarding release of novel organisms are:

1. Detection, identification and enumeration;
2. persistence in the environment;
3. fate and transport to niches other than those intended;
4. gene transfer to and from other microbes in the ecosystem; and
5. involvement in environmental effects
    a) pathogenicity and acute and chronic toxicity to non-target species
    b) disruption of environmental processes.

Test Methods Development for Risk Assessment of Novel Microbes 59

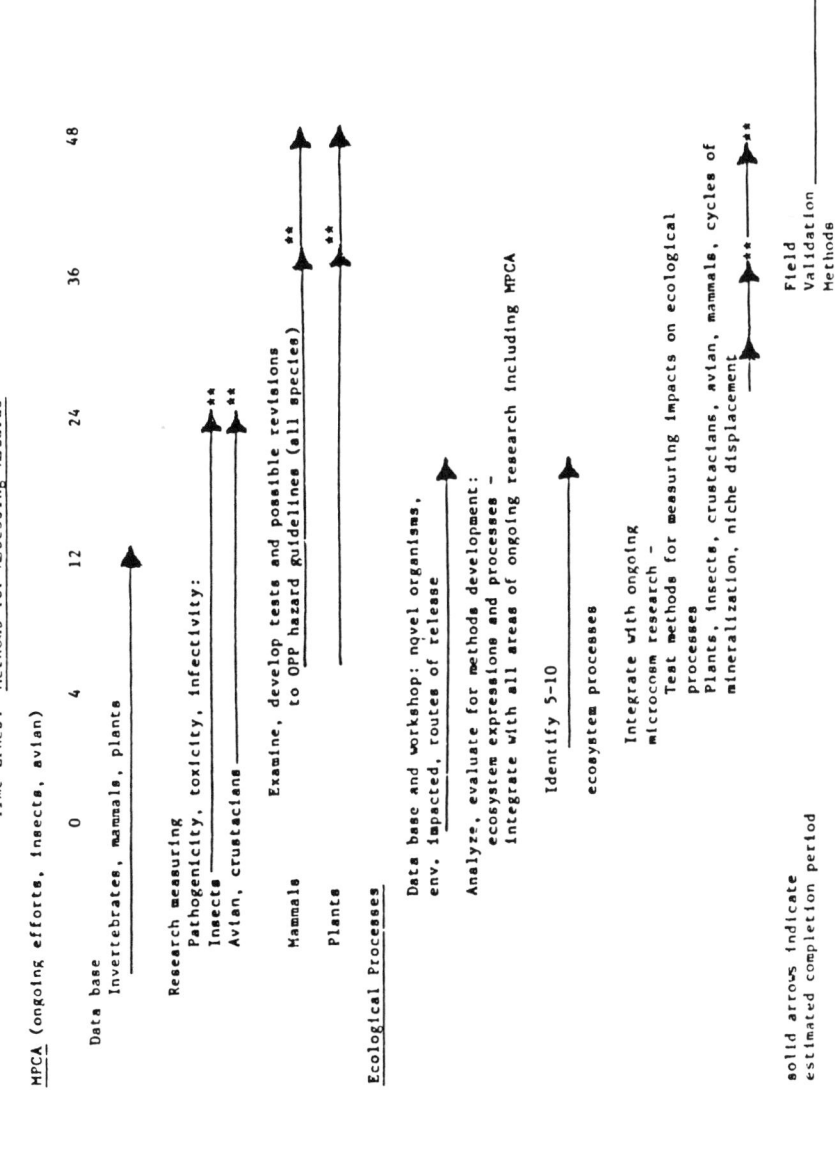

VIII. ACKNOWLEDGEMENTS

This research plan was written by Ramon J. Seidler (Oregon State University) with significant contributions from Jane Rissler (University of Maryland). Many individuals have influenced the form and composition of the final draft. Special appreciation is extended to Charles Hendricks, Al Bourquin, Guenther Stotzky, Bob Brink, and Fred Betz for their critical review and numerous editorial and technical suggestions. Appreciation is also extended to members of the AAAS Workshop for their reviews, suggestions, and encouragement. The coordination and development of the research plan is largely due to the efforts of Harold Kibby.

IX. LITERATURE CITED

1. Clay, D. 1983. Testimony before U.S. Congress House Subcommittee on Science, Research, and Technology and House Subcommittee on Investigations and Oversight. June 22, 1983. Washington, D.C.

2. OTS Research Needs: Background Material for the December ORD Biotechnology Workshop, OTS, EPA, Nov. 1983.

3. Biotechnology Workshop Report. 1983. Report on ORD-sponsored biotechnology workshop, December 14-15, 1983, Washington, D.C.

4. Lowrance, W. W. 1976. Of acceptable risk: science and the determination of safety. Wm. Kaufmann, Inc., Los Altos, CA, 180 pp.

5. National Research Council. 1983. Risk Assessment in the Federal Government: Managing the Process. National Academy of Science. National Academy Press, Washington, D.C., 191 pp.

6. Powledge, T. M. 1983. Public says genetic engineers should proceed cautiously. Biotechnology. October, pp. 645-646.

7. Powledge, T. M. 1983. Will restraints continue to loosen? Biotechnology. June, pp. 322-328.

8. Sharples, F. E. 1983. Spread of organisms with novel genotypes: thoughts from an ecological perspective. Recomb. DNA Tech. Bull. $\underline{6}$:43-56.

9. Milewski, E. 1983. Congressional hearing on the environmental implications of genetic engineering. Recomb. DNA Tech. Bull. $\underline{6}$:103-110.

10. Old, R. W., and S. B. Primrose. 1981. Principles of gene manipulation. University of California Press, Berkeley, California.

11. Clean-up microbes for the environment. Science News $\underline{119}$:246.

12. Kellogg, S. T., D. K. Chatterjee, and A. M. Chakrabarty. 1981. Plasmid-assisted molecular breeding: New technique for enhanced biodegradation of persistent toxic chemicals. Science $\underline{214}$:1133-1135.

13. Pemberton, J. M., and R. H. Don. 1981. Bacterial plasmids of agricultural and environmental importance. Agricul. Environ. 6:23-32.

14. Stotzky, G., and V. N. Krasovsky. 1981. Ecological factors that affect the survival, establishment, growth and genetic recombination of microbes in natural habitats. In Levy, S. B., C. Clowes, and E. L. Koenig (ed.) Molecular Biology, Pathogenicity, and Ecology of Bacterial Plasmids. Plenum Press, New York, NY.

15. Schwien, U. and E. Schmidt. 1982. Improved degradation of monochlorophenols by a constructed strain. Appl. Environ. Microbiol. 44:33-39.

16. Hardy, K. 1983. Bacterial plasmids. American Society for Microbiology. Washington, D.C.

17. Boucher, C., B. Message, D. Debieu, and C. Zischek. 1981. Use of P-1 incompatability group plasmids to introduce transposans into Pseudomonas solanacaearum. Phytopathol. 71:639-642.

18. Ditta, G., S. Stanfield, D. Corbin, and D. R. Helinski. 1980. Broad host range DNA cloning system for Gram-negative bacteria: Construction of a gene bank of Rhizobium meliloti. Proc. Natl. Acad. Sci. USA 77:7347-7351.

19. Barth, P. T., L. Tobin, and G. S. Sharpe. 1981. Development of broad host-range plasmid vectors. In Levy, S. B., C. Clowes, and E. L. Koenig (ed.) Molecular Biology, Pathogenicity, and Ecology of Bacterial Plasmids. Plenum Press, New York, NY.

20. Kleckner, N., R. K. Chan, B. K. Tye, an D. Botstein. 1975. Mutagenesis by insertion of drug resistance element carrying an inverted repetition. J. Mol. Biol. 97:561-575.

21. Leemans, J., D. Inzé R. Villarroel, G. Engler, J. P. Hernalsteens, M. DeBlock and M. Van Montagu. 1981. Plasmid mobilization as a tool for *in vivo* genetic engineering. *In* Levy, S. B., C. Cloews, and E. L. Koenig (ed.) Molecular Biology, Pathogenicity, and Ecology of Bacterial Plasmids. Plenum Press, New York, NY.

22. Kirk, T. K. 1983. Biotechnology in utilization of wood. Biotechnology. October, pp. 666-668.

23. Beringer, J. E. 1979. The development of *Rhizobium* genetics. J. Gen. Microbiol. 116:1-7.

24. Levine, M. M., J. B. Kaper, H. Lockman, R. E. Black, and M. L. Clements. 1983. Recombinant DNA risk assessment studies in man: Efficacy of poorly mobilizable plasmids in biologic containment. Recomb. DNA Tech. Bull. 6:89-97.

25. Terrestrial Microcosms and Environmental Chemistry. 1977. *In* Witt, J. M., and J. W. Gillet (eds.). The Proceedings of Two Colloquia, June 13-14, 1977, OSU, Corvallis, OR. National Science Foundation, Washington, D.C. 20550.

26. Terrestrial Microorganisms. 1977. *In* Gillet, J. W. and J. M. Witt. (eds.). The Proceedings of the Workshop of Terrestrial Microcosms. National Science Foundation, Washington, D.C. 20550.

27. Personal communication, Dr. Craig McFarlane. Corvallis Environmental Research Laboratory, Toxic and Hazardous Materials Branch, U.S. EPA, Corvallis, Oregon 97331.

28. Kleeberger, A., and W. Klingmüller. 1980. Plasmid-mediated transfer of nitrogen-fixing capability to bacteria from the rhizosphere of grasses. Molec. Gen. Genet. 180:621-627.

29. Kralikova, K., V. Krcmery, Sr., and V. Krcmery, Jr. 1983. Incidence of gentamicin-resistant bacteria and those with R. plasmids, transferable to E. coli K-12, in municipal waste waters. Recomb. DNA Tech. Bull. 6:98-100.

30. Schilf, W., and W. Klingmüller. 1983. Experiments with Escherichia coli on the dispersal of plasmids in environmental samples. Recomb. DNA Tech. Bull. 6:101-102.

31. Talbot, H. W., Jr., D. K. Yamamoto, M. W. Smith, and R. J. Seidler. 1980. Antibiotic resistance and its transfer among clinical and nonclinical Klebsiella in botanical environments. Appl. Environ. Microbiol. 39:97-104.

32. May, S. N., and B. B. Bohlool. 1983. Competition among Rhizobium leguminosarum strains for nodulation of lentils (Lens esculenta). Appl. Environ. Microbiol. 45:960-965.

33. Kosslak, R. M., B. B. Bohlool, S. Dowdle, and M. J. Sadowsky. 1983. Competition of Rhizobium meliloti strains in early stages of soybean nodulation. Appl. Environ. Microbiol. 46:870-873.

34. Yamamoto, D. K., and R. J. Seidler. 1981. Colonization of growing radish plants by clinical and nonclinical isolates of Klebsiella inoculated onto seeds. Curr. Microbiol. 5:289-293.

35. Kloepper, J. W., M. N. Schroth, and T. D. Miller. 1980. Effects of rhizosphere colonization by plant growth-promoting rhizobacteria on potato plant development and yield. Phytopathol. 70:1078-1082.

36. Schroth, M. N., and J. G. Hancock. 1982. Disease-suppressive soil and root-colonizing bacteria. Science 216:1376-1381.

37. Johnston, A. W. B., G. Hombrecher, and N. J. Brewin. 1981. Rhizobium plasmids: their role in the nodulation of legumes. In Levy, S. B., C. Clowes, and E. L. Koenig (ed.) Molecular Biology, Pathogenicity, and Ecology of Bacterial Plasmids. Plenum Press, New York, NY.

38. Johnston, S. W. B., J. L. Beynon, A. V. Buchanan-Wollaston, S. M. Setchell, P. R. Hirsch, and J. E. Beringer. 1978. High frequency transfer of nodulating ability between strains and species of Rhizobium. Nature 276:634-636.

39. Suslow, T. V., and M. N. Schroth. 1982. Rhizobacteria of sugar beets: Effects of seed application and root colonization on yield. Phytopathol. 72:199-206.

40. Suslow, T. V., and M. N. Schroth. 1982. Role of deleterious Rhizobacteria as minor pathogens in reducing crop growth. Phytopathol. 72:111-115.

41. Kloepper, J. W., and M. N. Schroth. 1981. Relationship of in vitro antibiotics of plant growth-promoting rhizobacteria to plant growth and the displacement of root microflora. Phytopathol. 71:1020-1024.

42. Broadbent, P., K. F. Baker, and Y. Waterworth. 1971. Bacteria and actinomycetes antagonistic to fungal root pathogens in Australian soils. Aust. J. Biol. Sci. 24:925-944.

43. Terrestrial Microcosms. 1977. Proceedings of a workshop on terrestrial microcosms. J. W. Gillett and J. M. Witt (ed.). National Science Foundation, Washington, D.C.

44. Torsvik, V. L., nad J. Goksoyr. 1977. Determination of bacterial DNA in soil. Soil. Biol. Biochem. 10:7-12.

45. Torsvik, V. L. 1983. Isolation of bacterial DNA from soil. Soil. Biol. Biochem. 12:15-21.

46. Don, R. H., and J. M. Pemberton. 1981. Properties of six pesticide degradation plasmids isolated from Alcaligenes paradoxus and Alcaligenes eutrophus. J. Bacteriol. 145:681-686.

47. Rissler, J. F. 1983. Research needs for biotic environmental effects of genetically-engineered microorganisms. Draft document.

48. Federal Register. 1983. Guidelines for research involving recombinant DNA molecules. 48:24556-24581.

49. Gerhardt, Phillipp, Editor-in-Chief. 1981. Manual of methods for general bacteriology. American Society for Microbiology, Washington, D.C.

50. Collins, C. H., and Patricia M. Lyne. 1976. Microbiological Methods. Butterworth & Co. London, England.

51. Aaronson, Sheldon. 1970. Experimental Microbial Ecology. Academic Press, Inc. New York, NY.

52. The Prokaryotes. 1981. A handbook on habitats, isolation, and identification of bacteria. Volumes I and II. Springer-Verlag, New York.

53. Meyers, J. A., Sanchez, D., Elwell, L. P., and Falkow, S. 1976. Simple agarose gel electrophoresis method for the identification and characterization of plasmid deoxyribonucleic acid. J. Bacteriol. 127:1529-1537.

54. Grunstein, M., and Hogness, D. S. 1975. Colony hybridization: a method for the isolation of cloned DNAs that contain a specific gene. Proc. Natl. Acad. Sci. USA 72:3961-3965.

55. Moseley, S. L., Huq, I., Alim, A. R. M. A., So., M., Samadpour-Motalebi, M., and Falkow, S. 1980. Detection of enterotoxigenic Escherichia coli by DNA colony hybridization. J. Infect. Dis. 142:892-898.

56. Masui, Y., T. Mizuno, and M. Inouye. 1984. Novel high-level expression cloning vehicles: 10^4-fold amplification of Escherichia coli minor protein. Biotech. 2:81-85.

57. Kohn, G. K. 1980. Bioassay as a monitoring tool. Residue Rev. 76:99-127.

58. Mulla, M. S., G. Majori, and A. A. Arata. 1979. Impact of biological and chemical mosquito control agents on nontarget biota in aquatic ecosystems. Residue Rev. 71:121-173.

59. Bourquin, A. W., P. H. Pritchard, R. Vanolinda, and J. Samela. 1979. Fate of Dimilin in Estuarine microcosms. Absts. Ann. Meeting. American Society of Microbiology, Q76, page 232.

60. Pritchard, P. H., A. W. Bourquin, H. L. Frederickson, and T. Mazianz. 1979. System design factors affecting environmental fate studies in microcosms. In: Proc. Workshop:microbial degradation of pollutants in marine environments. US EPA, Gulf Breeze, FL. pp. 251-272.

61. Bourquin, A. W. 1977. Effects of malathion on microorganisms of an artificial salt-marsh environment. J. Env. Qual. 6:373-378.

62. Liang, L. N., J. L. Sinclair, L. M. Mallory, and M. Alexander. 1982. Fate in model ecosystems of microbial species of potential use in genetic engineering. Appl. Environ. Microbiol. 44:708-714.

63. U.S. E.P.A., 1983. Pesticides Assessment Guidelines, Subdivision M - Biorational Pesticides. National Technical Information Service, Springfield, VA. No. PB83-153965.

64. Betz, F., M. Levin, and M. Rogul. 1983. Safety aspects of genetically-engineered microbial pesticides. Recomb. DNA Tech. Bull. 6:135-141.

65. Camann, D. E. 1980. A model for predicting dispersion of microorganisms in wastewater aerosols. pp. 46-70. In Wastewater aerosols and disease. U.S. EPA 600/9-80-028.

BIOTECHNOLOGY HEALTH RISK ASSESSMENT RESEARCH PLAN

Marvin Rogul
The Rogul Group
Washington, DC

John R. Fowle III
Office of Health Research, EPA
Washington, DC

I. INTRODUCTION

Certain products of biotechnology fall within the regulatory purview of the Office of Pesticides and Toxic Substances (OPTS) under the jurisdiction of the Toxic Substances Control Act (TSCA) and the Federal Insecticide, Fungicide and Rodenticide Act (FIFRA). In order to accomplish their mission under these Acts, OPTS requested that the Office of Research and Development (ORD) provide technical assistance in the development of models and data bases for use to evaluate biotechnology products. This chapter presents a research approach developed by the Health Effects Workgroup (HEWG) of the Coolfont Workshop for using data and applying test methods to assess the possibility of health risks associated with exposure to genetically engineered microbes released into the environment. New test methods may be developed as the need arises. This chapter presents a research plan, not a research proposal. Specific proposals will be developed from this plan. In all cases EPA will base its activities on the extensive body of information concerning medical microbiology and will coordinate work with other Agencies with health related missions, such as the National Institutes of Health (NIH) and the Food and Drug Administration (FDA).

The Coolfont Health Effects Work Group (HEWG) recommendations are described in this chapter with discussions of certain considerations about risk assessment that were factored into the development of the research plan. The key role that the NIH Recombinant DNA Advisory Committee (RAC) has played in assessing risks from certain genetically engineered organisms and in providing guidance for future efforts is recognized at the conclusion of the chapter.

This brief review points out the nature of the work EPA intends to pursue with respect to the possible health effects of genetically engineered organisms released into the environment.

II. HEALTH EFFECTS WORK GROUP PANEL RECOMMENDATIONS

The major conclusion of the Coolfont HEWG was that models and test methods to predict the potential health effects arising from the accidental or deliberate release of biotechnology products are not scientifically feasible at this time because of the variety of potential applications of biotechnology and the types of organisms which could be used in commerce. The panel emphasized that because of this, health assessments of biotechnology products must be performed on a case-by-case basis.

Two major areas were identified where useful research could be conducted: 1) the development of animal testing protocols to determine the health effects of genetically altered organisms and 2) specific experiments on agents of immediate concern to the Agency. The conclusions and general recommendations provided by the HEWG are found in Table 1 and are discussed below.

A. Data Gathering and Information Management

It was stressed that steps should be taken to insure access to a wide range of pertinent, up to date information, including relevant computer data bases, use of computers to catalog in-house experience, storage of pertinent DNA sequence information, information about tissue banks and monitoring data, etc.

As part of the emphasis to build upon past experience HEWG emphatically recommended that tests be developed in accordance with the principles contained in the Pesticide Assessment Guidelines of Subdivision M: Biorational Pesticides which were developed by the EPA in 1982.[1] (See Table 2 for an outline of the Subpart M Testing Approach.)

Table 1. Conclusions and General Recommendations

1. Health research on the pathogenic potential of genetically engineered microorganisms should be directed at answering questions for specific organisms and specific situations. General situations do not presently exist. An accumulation of data from many individual cases may eventually be used to generate more general guidelines.

2. Since the prediction of potential health effects from genetically engineered organisms is not completely attainable or scientifically feasible in many cases, research aimed at developing generic and specific models and test methods should be conducted in this area. For the present, decisions on hazard and risk potential must be made on a case-by-case basis.

3. Development of information systems may be useful for literature searches and the cataloging of data on biotechnology, rather than as an aid to predicting pathogenicity.

4. EPA must have ready access to existing literature on microbial pathogenicity and biotechnology. An extensive library collection should be developed.

5. The Agency has to develop in-house expertise and strong communications with non-agency experts to review the health effects potential of new microorganisms. Optimally, in-house experts should be centralized in one location.

6. EPA projects in biotechnology should go through rigorous peer review similar to that performed at NIH. It is recommended that the Agency convene a special subpanel of the Science Advisory Board and the Science Advisory Panel to act as a review committee and that each laboratory adopt an internal review mechanism and Institutional Biosafety Committee.

7. Subdivision M guidelines for microbial pest control agents should be reexamined and revised where necessary. The testing of viral agents should be expanded and the guidelines should be validated with known microorganisms (pathogenic and nonpathogenic).

8. To consider the potential pathogenicity of a new microorganism, it is essential to first consider the nature of the source organisms and the construct. If health effects testing is perceived to be necessary, it must be performed in animal models which closely approximate human experience. If data on human health effects are available, the data should be given very high consideration when developing regulatory decisions.

9. It is recommended that the advice of a team of experts from certain key disciplines be made available to manufacturers at early stages in the development of products. The team of experts would work with manufacturers to assess possible risks and to suggest options and appropriate tests prior to commercialization of a genetically engineered organism. The team would include both EPA and outside experts.

10. It is recommended that funds be allocated to the Office of Research and Development for the development of a continuing program to conceive, develop, and evaluate the use of a variety of genetic markers for use in tagging microorganisms (for identification purposes) and to advise industry in their use. Genetic markers might include unusual resistance patterns to antibiotics not used in human or animal medicine, unusual fluorescence and metal ion resistance, or production of unique proteins. Thus, tagged organisms could be identified or dismissed as etiological agents of disease.

Table 2. Outline of FIFRA Subpart M Microbial Pesticide Control Agent Test Requirements

Tier I tests for Microbial Pest Control Agents:

1. Acute oral toxicity/infectivity in the rat (LD_{50})
2. Acute dermal toxicity/infectivity, rat or mouse (LD_{50})
3. Acute inhalation toxicity/infectivity in the mouse, rabbit or guinea pig (LD_{50})
4. Intravenous intracerebral, and intraperitoneal toxicity/infectivity using rabbits, new born or newly weaned mice and hamsters.
5. Primary dermal irritation on guinea pigs or rabbits
6. Primary eye irritation on rabbits
7. Hypersensitivity in hamsters or rabbits
8. Hypersensitivity incidents
9. Cellular immune responses in mice
10. Tissue culture tests for viruses

Tier II tests include:

The tests in this Tier are similar to Tier I but somewhat more extensive in that they add a subchronic oral test, teratogenicity, mutagenicity, and virulence enhancement tests. Dogs are added as test animals and the number of animals used are increased.

Tier III tests:

1. Chronic oral test in the rat
2. Oncogenicity test in newly weaned mice and rats
3. Mutagenicity test in mammals
4. Teratogenicity test in 2 species from rat, mouse, hamster or rabbit

B. Selection of Organisms for Validating Subpart M Test Approach

The HEWG advised that a few key organisms should be studied using the approach outlined by the Subpart M guidelines to determine its validity. They suggested that the organisms developed to mitigate environmental pollutants in EPA's Engineering Research Laboratories should be included in this effort. The group suggested that Pseudomonas syringae "ice minus" and Pseudomonas fluorescens containing the B. thuringiensis toxin gene would be good candidates for study also.

C. Protocol Development for Infectivity, Pathogenicity, and Metabolic Characteristics of Recombinant microorganisms

The HEWG recommended that EPA develop and validate more animal pathogenicity test protocols and improve the testing component for viruses, viral products, and, if appropriate, oncogenes in the Subdivision M guidelines. The workgroup believed that a health testing scheme following the tiered testing methods of the Subdivision M guidelines would provide the Environmental Protection Agency with experience that might enable the development of more specific and appropriate tests. Such tests would employ non-human animal models or would test certain parameters of importance to human physiology and could be used to demonstrate the reasonable safeness of genetically engineered products and GEMS prior to EPA approval for commercial production and application. One such test could be the ability to grow at 37°C. This test would distinguish organisms which could grow systemically at human body temperature. It would be reasonable to assume that most organisms which could not grow at this temperature would not be systemic pathogens. The production of toxic effects in human tissue culture and the production of exo- or endo-toxins (such as toxin A or elastase in Pseudomonads, exotoxins in Corynebacteria) or invasins (such as coagulase and laminin in Staphylococci) or oncogenic material could also be added to Tier I testing protocols for the purpose of excluding potentially harmful organisms from being released to the environment with subsequent human exposure.

D. Bacterial Pathogenicity Categories

The Agency was advised to compile a list of the characteristics of pathogens and nonpathogens that would have commercial application and to advise industry on its use. The Agency was advised to develop bacterial pathogenicity categories ranging from known pathogens to organisms incapable of infecting humans, and if faced with the situation, to place special emphasis on the microorganisms that might be constructed from oncogenic viruses. It was recommended that the Agency support the evaluation and validation of the Subdivision M pesticide guidelines. This should be done for genetically engineered microbes, non-GEMS and a variety of viruses and vectors. Expert advice should be sought from outside the Agency to guide these efforts.

From the environmental health viewpoint, most risk assessment work has been done with *E. coli* and is of limited value from the environmental standpoint. There is little information on other organisms which are being considered for commercial use in large scale contained facilities or for release to the environment.

Except for allergies, and the rare infection due to accidential exposure to the frank pathogens used in the production of toxins and antigens for vaccines, there are no major problems in the traditional biotechnology industries of baking, wine and beer making or other commercial activities using microorganisms. A number of companies (Sybron Biochemical, Polybac and Flow Laboratories) have had at least a decade of experience in degrading environmental waste by using microorganisms in bioaugmentation programs. No known health mishaps such as infection have occurred.[2] Although this does not guarantee that there will be no human health hazards from environmental release of genetically engineered organisms, it does show that our past experience with the commercial use of microbes has not resulted in an unreasonable risk. On this basis the HEWG opined that releases of genetically engineered microbes are not likely to pose a health risk to humans unless they are pathogens, related to pathogens, or contain a gene whose product is toxic to humans. The large number of possible variations precludes a predictive model or the ability to test each and every product. Thus, it was recommended that each product be evaluated on a case-by-case basis and that advantage be taken of the experience gained from each evaluation to improve the Agency's risk assessment capabilities. It is possible that relevant health information will come from workers in the laboratory and at commercial production facilities, where products of genetic engineering are developed and made. Previous epidemiology studies of populations exposed to microbial agents (e.g., sewage treatment plant workers, populations exposed to wastewater aerosols) should be critically reviewed to determine their relevancy to determining risks from exposure to GEMS. (Retrospective epidemiology studies could be performed on workers who have been working in traditional fermentation and other biotechnology firms. Prospective studies could also be performed on these firms, the new genetic engineering firms and biological pesticide applicators.)

E. Establishment and Management of a Data Base of Characteristics of the Potential Hazards of Genetically Modified Materials

Genetically engineered organisms should be evaluated using existing data on parent, vector and host organisms and if needed, animal tests. This should be used to support recommendation 1 above on infectivity, pathogenicity and metabolic characteristics of recombinant microbial organisms.

F. Selection and Assessment of Safe Hosts

HEWG recommended that the EPA identify microorganisms that are considered to be relatively safe for use in developing new organisms and that these "source" or host organisms be tested using the Subdivision M guidelines.

G. Development of Molecular Probes

The development of DNA/DNA, DNA/RNA, and RNA/RNA hybridization probes as well as serological and other immunological tests are essential for diagnosis and monitoring of interactions between genetically engineered organisms and other species. These tests and their modifications, as well as conventional monitoring techniques, are discussed in the monitoring chapter of this report. The HEWG felt that the development and application of such techniques are essential to any program evaluating potential health effects from GEMS. One advantage of these approaches is that the probes can specifically detect nucleotide sequences or gene products. A new generation of avidin [3,4] complexing biotin probes holds special promise for the future. The development of such approaches was thought to have potential as an excellent diagnostic tool, and baculoviruses were recommended as valuable research models for the development and application of such probes because of the use of baculovirus for pesticide control purposes. Such work would have immediate benefit for EPA decision-making.

III. DISCUSSION

It was stressed by the HEWG that the use of animal surrogates is not highly effective in detecting possible human opportunists or even pathogens. That is why validation of these tests is extremely important.

A. Risk Assessment

To perform a risk assessment, data on (1) hazard identification and characterization, (2) exposure and (3) dose/response relationships are required. In the case of microorganisms it is assumed that after a release to the environment risk is dependent on the survival of the organism, and its growth and reproduction in a receptive environment; the consequences can be either beneficial or deleterious. According to Dr. Martin Alexander the construction of new genotypes and genetic exchanges in the environment may lead to unexpected phenotypes and functions, and because there is very little known about these processes, there is not a sufficient data base from which to make adequate risk assessments.[5]

Prior to commercial production of GEMS, and their release to the environment, some evaluation of potential health effects should be performed based on existing literature or specific testing. The following characteristics would be useful in assessing risks in experimental animals and attempting to develop and validate extrapolation on approaches to human risk. (The assumption in this line of reasoning is that generally acceptable experimental animals models can be developed):

A. Nature and degree of pathogenicity/toxicity
  1. Pathogenicity and virulence of viable organisms
     a. route of infection
     b. invasiveness in host

c. replication in host
2. Toxicity due to viable or nonviable organisms or products
   a. acute or chronic
   b. mutagenicity
   c. teratogenicity
   d. oncogenicity
   e. immune effects (toxicity and hypersensitivity)

B. Presence of other intracellular or extracellular agents (e.g. viruses) in the product.

C. Degree of debility
   1. auxotrophy
   2. antibiotic sensitivity/dependence
   3. pH sensitivity, light, temperature, (other physical factors)

D. Genetic stability
   1. Under the conditions which it is to be used
      a. environment (factory, outdoors)
      b. host

E. Influence of the vector/insert on construct stability

F. State of the vector in the cell (integrated or non-integrated)

G. Mobility of vector

H. Expression of functions

I. Infectivity to other humans and animals

In genetically engineered microorganism there are two special concerns which may relate directly or indirectly to human health. One concern will be of genetic interchange between introduced organisms and those which normally inhabit humans. The other consideration is the possible interchange between DNA from GEMS and human DNA.

Special consideration must be given to viruses, because they can integrate into mammalian DNA and special probes should be made to identify viral sequences if these viruses are to be used for commercial purposes and released to the envirnment.

B. Foundation Laid by the NIH Recombinant DNA Advisory Committee

The members of the Recombinant DNA Advisory Committee (RAC) of the National Institutes of Health (NIH) have had extensive experience with risk assessment of genetically engineered organsisms. Although most of the RAC experience has been medically oriented and limited to work conducted with recombinant DNA (R-DNA) techniques to the exclusion of all other forms of genetic engineering and manipulation, such as cell fusion, transduction, transformation, conjugation and mutation, organizations such as the EPA can benefit from their approach and experience.

One of the major activities of the RAC was to formulate the NIH Guidelines for Research Involving Recombinant DNA Molecules. The guidelines have designated certain organisms as generally safe to work with in the laboratory and usually exempt them from experimental guidelines. The organisms are the attenuated Escherichia coli strain K-12, an asporogenic Bacillus subtilis, and the yeast Saccharomyces cerevisiae. These are very well characterized organisms and generally believe to be safe with little chance of escaping from the laboratory; incapable of surviving for any great length of time outside a laboratory environment; or incapable of causing harm to humans or the environment.

In fact, NIH has funded a great deal of work to characterize the potential human health risks of the medically important bacterium Escherichia coli, the major bacterium for basic research in bacterial genetic engineering. Their approach was discussed by the HEWG as a useful model for future efforts.

1. **E. coli Studies which Influenced the Development of the RAC Guidelines**

Experiments were performed to determine whether certain strains of debilitated E. coli strain K-12, could: (1) survive outside of the laboratory, (2) survive in the mammalian gastrointestinal system, and (3) transfer genetic information between strains in the mammalian gut. The results indicated that E. coli was a safe organism to use in that although E. coli K-12 could survive in sewage, it could not persist in the human alimentary system.[6] In fact, when laboratory personnel who worked with E. coli K-12 and R-plasmids (resistance factors; also called R factors) were screened to determine whether their endogenous enteric flora were contaminated with K-12 or R factors the results were negative. There was no evidence of infection or R factor persistence in the gut.[7]

In human experiments greater than 10^9 E. coli strain K-12 bacteria were fed to volunteers. And although some of the organisms survived passage through the digestive system, none of the organisms persisted in the intestinal tract for more than a week or so, even under conditions of extreme antibiotic pressure that would have been expected to aid establishment in the gut.[8,9,10,11,12] It was also found that E. coli K-12 was not capable of transfering poorly mobilized plasmid vectors such as pBR325 in the presence of F-amp (fertility mobilizing factor) in human volunteer studies. In fact, even a colonizing strain such as E. coli HS could not transfer pBR325 unless antibiotic pressure somehow forced and selected for the transferred vector in resident E. coli. However, E. coli HS could participate in triparental conjugation in the human intestine in the presence of tetracycline, and under these conditions transfer the non-conjugative pJBK5 to the resident E. coli.[12]

From these findings it is evident that when microorganisms are specifically debilitated and properly constructed, it is very difficult to promote human infection and genetic interchange among them. The exception appears to be when highly selective artificial environmental conditions, such as antibiotic pressure are imposed upon the microorganisms.

Other organisms are also characterized by the RAC guidelines, but not exempted from the guideline recommendations. These organisms are usually characterized by their host characteristics and the genetic vectors they contain. These host-vector (HV) systems are in turn assigned biosafety levels comparable to their perceived level of risk. The greater the danger of the host-vector, the higher the level of containment in the biosafety system assigned.

2. Experiments Simulating High Risk Conditions: Promoting and Detecting Genetic Interchange

Experiments sponsored by NIH have been designed to simulate high risk conditions. In two of the most well known series of experiments, polyoma tumor virus DNA was cloned into strains of E. coli K-12 with both plasmid and phage vectors. In various forms the cloned DNAs were introduced into mice or cultured mouse cells. Evidence of polyoma virus infection was to be assayed by measuring the immunological response to polyoma proteins. Because of the potential dangers, these risk experiments were carried out at the highest level of physical containment.

The results were consistent with the belief that these procedures are extremely unlikely to produce evolutionarily fit, epidemic pathogens. There was no immunological evidence for polyoma infection in mice when the polyoma DNA was borne by a plasmid or by E. coli that were lysogenic for lambda-polyoma prophage.

On the other hand, in this same series of experiments, immunological evidence for biological activity was observed when dimers of the polyoma DNA were carried by free-phage, and when the total DNA of the E. coli/ lambda-polyoma chimeras was introduced into the mouse.[13][14] However, recent work has shown that prolonged feeding or inoculation of wild type of laboratory strains of E. coli containing monomeric or dimeric forms of polyoma virus to conventional, antibiotic-compromised and germ-free mice did not demonstrate infection of the mice by polyoma virus.[15] On the basis of these findings it would be prudent to be especially mindful of unexpected molecular interactions when dealing with any kind of mammalian and retrovirus DNA. Preliminary data from EPA investigations in tissue cultures warrant a close look at insect viral interactions with human DNA (Kawanishi, C. personal communication).

IV. REFERENCES

1. Betz FS, Beusch WR, Brittin EB, Carsel R, Cohen SZ, Holst RW, Keller A, Mauer IN, Roessler W, Urban D, Vaughan A, and Woodrow W. Pesticide Assessment Guidelines: Subdivision M, Biorational Pesticides. National Technical Information Service. Springfield Va. 1982.

2. Genetic Control of Environmental Pollutants. ed. Omenn GS and Hollaender A. pp. 331-349. Plenum Press, New York. 1984.

3. Lewin R. Gene Probes Become Ever Sharper. Science 221:1167, 1983.

4. Richman D et al. Summary of a Workshop on New and Useful Methods in Rapid Viral Diagnosis. J Infect Dis. 150:941-951, 1984.

5. Alexander, M. Spread of Organisms with Novel Genotypes in Biotechnology and the Environment: Risk and Regulation. (eds) A. Teich, MA Levin and JH Pace. AAAS, Washington, DC. 1985.

6. Sagik BP and Sorber CA. The Survival of Host-Vector Systems in Domestic Sewage Treatment Plants. Rec DNA Tech Bull 2:55-61, 1979.

7. Petrocheilou V, and Richmond MH. Absence of plasmid of Escherichia coli K-12 infections among laboratory personnel engaged in R-plasmid research. Gene 2:323-327, 1977.

8. Smith HW. Survival of orally administered E. coli K-12 in the alimentary tract of man. Nature 255:500-502, 1975.

9. Anderson ES. Viability of and transfer of a plasmid from E. coli K-12 in the human intestine. Nature 255:502-504, 1975.

10. Formal SB, Hornick RB. Invasive Escherichia coli. J Infect. Dis 137:641-644, 1978.

11. Levy SB, Marshall B, Rowse-Eagle D. Survival of Escherichia coli host-vector systems in the mammalian intestine. Science 209:391-394, 1980.

12. Levine MM, Kaper JB, Lockman H, Black RE, Clements ML and Falkow S. Recombinant DNA Risk Assessment Studies in Man: Efficacy of Poorly Mobilizable Plasmids in Biologic Containment. Recombinant DNA Technical Bulletin 6:89-97, 1983.

13. Israel, MA, Chan HW, Rowe WP and Martin MA. Molecular cloning of polyoma virus DNA in Escherichia coli: plasmid vector system. Science 203:883-887, 1979.

14. Chan HW, Israel MA, Garon CF, Rowe WP, and Martin MA. Molecular cloning of polyoma virus DNA in Escherichia coli: lambda phage vector system. Science 203:887-892, 1979

15. Smith, C. Jr, E Milewski and MA Martin. The effects of colonizing mice with laboratory and wild type strains of Escherichia coli containing tumor virus genomes. Recombinant DNA Technical Bulletin 8: 47-51, 1985.

ENVIRONMENTAL ENGINEERING RESEARCH SUPPORT PROPOSAL

John Burckle
Industrial Environmental Research Lab., EPA
Cincinnati, Ohio

Albert D. Venosa
Environmental Research Lab., EPA
Cincinnati, Ohio

I. LEGISLATION

The Toxic Substances Control Act (TSCA) mandates the regulation of chemical substances, new or existing, which present an "unreasonable risk of injury to health or the environment". Section 5 of TSCA requires that any person who intends to manufacture a new substance other than for Research and Development, must submit a notice containing certain information to EPA for review at least 90 days before manufacture. The Office of Toxic Substances (OTS) has the authority to prevent manufacture, regulate use, or permit unregulated use depending upon the findings of this review.

II. REGULATORY NEEDS

Because genetically engineered microorganisms (GEMS) have been determined to constitute a "new chemical" by the Office of General Counsel of the Environmental Protection Agency (EPA), the OTS required to conduct reviews of Premanufacture Notices (PMN) for such substances. To conduct such PMN reviews, the OTS needs appropriate technical information and predictive capabilities to assess the potential risks arising from the release of and exposure to these substances, and to evaluate proposed alternatives and the costs for preventing release and exposure.

In preparation for the ORD (Office of Research and Development) Biotechnology Workshop I, (December 1983), OTS issued extensive material for the purpose of determining specific research needed in support of anticipated OTS regulatory activities. Subsequently, the regulatory needs of OTS were stated in the draft document, "EPA's Proposed Approach for Providing Technical Support in Biotechnology and Meeting the Research Needs of the Program Offices". Evaluations of PMNs are to be the "primary thrust" of OTS activity. Emphasis is to be placed on identifying existing data bases and expanding capabilities for predictive risk assessment purposes. In addition, the Workshop I specifically identified the objective and scope of the Workshop II (May 1984).

The focus of the workshop and hence this proposal, reflects OTS's primary concern regarding deliberate and accidental release into the environment and subsequent exposure of humans and the environment to such substances. Genetically engineered microorganisms used as pesticides, drugs, cosmetics, and food are specifically excluded from TSCA jurisdiction and therefore are not addressed here. However, genetically engineered microorganisms "manufactured" for the production of substances so used may be of concern and are addressed in the context of industrial manufacturing processes and plants.

Further, OTS has limited its concern, for the purposes of this workshop, to genetically modified microorganisms that are viruses, bacteria, fungi, algae, or protozoa. It is not within the scope to discuss the risks associated with (1) accidental environmental release or worker exposure to the vectors (e.g., plasmids, episomes, viroids); or (2) exposure or release of commercial products produced by genetically modified organisms (e.g., enzymes used as catalysts). Chemical byproducts and contaminants (e.g., toxins, companion viruses) which are not intentionally produced by the manufacturer as a commercial product but are associated with the genetically modified organism are to be considered in workshop discussions because the risks due to by-products and contaminants may be considered to be additional risks associated with the genetically modified organism itself.

In addressing these areas, the OTS has posed the following information needs in the previous workshop.

1. Are there other concerns regarding control technologies, industrial release, and worker exposure?
2. Are there concerns specific to any of the following modified organisms: viruses, bacteria, fungi, algae, and protozoa?
3. For each of the above concerns, what information (in addition to the following) will OTS need to evaluate its concerns?
   a. identification of production processes, equipment, and practices which may result in worker exposure or environmental release;
   b. identification of control methods available to prevent or reduce exposure/release for each media.
4. Are there predictive tools or test methods available for obtaining the information needed to evaluate each concern?
5. If predictive tools or test methods are lacking, can ORD develop predictive tools or test methods? If so, ORD should describe the proposed research project in depth, including: objectives, limitations, research methods, the end product of the project, and how it will specifically address OTS needs; how long the project will take to complete; and projected costs.
6. Are there certain manufacturing techniques or uses to which OTS should pay particular attention?

III. OVERALL PROGRAM APPROACH

The overall risk assessment model is envisioned as composed of a series of modules which, when linked together, will permit estimation of the risk associated with the production and subsequent use in commerce of specific genetically engineered microorganisms. These modules, likely to be composed of a number of submodules, address the following aspects:

- engineering - potential for accidental and deliberate release and prevention and control of releases
- environmental - potential for survival in the environment and adverse effects
- health - potential for adverse health effects
- monitoring - techniques for detection and strategies for routine monitoring to ensure that release is not occurring, and, in the event of a release, that timely remedial actions can be taken

While limitations exist in the technical information data base required for the preparation of the assessment modules, they are more related to lack of knowledge of the potential behavior and characteristics of the specific genetically engineered microorganisms to be evaluated. This aspect is addressed by a recent congressional report (Environmental Implications of Genetic Engineering - Staff Report, Committee on Science and Technology U.S House of Representatives, 98th Congress, 1984) which examines the issues of regulation. This report acknowledges the "newness" of this area and states that..."Many of the standard approaches to the review of "conventional" chemical substances would not be applicable". The problem is that the effects of the substances cannot be adequately predicted in advance because there is no accumulation of knowledge to permit such prediction as there is with chemicals based upon structure activity analyses. The report also points out that the OTS chemical staff conducting PMN reviews would require ..."familiarity with containment practices for pathological organisms and the associated worker protection techniques...".

Because the physical, chemical, and biological aspects of the environment affecting the viability, behavior, and interactions of genetically engineered microorganisms in that environment (e.g., temperature, moisture, accessibility and level of nutrients, oxygen, predators, pathogens, etc.) and the specific properties resulting from the genetic modifications cannot be specifically anticipated in advance, these must be determined through the health and environmental effects assessment procedures. Once these data are developed, they can be used in assessment protocols to estimate risks.

IV. SUMMARY OF PROPOSED ENVIRONMENTAL ENGINEERING EFFORTS RELATED TO REGULATORY NEEDS

A. Regulatory Needs

The OTS has expressed three major concerns in regard to environmental engineering technology:

1. accidental or deliberate release of the genetically engineered microorganism from the site of production (e.g., in effluents), during transport (shipping), intermediate storage and subsequent use;

2. availability and effectiveness of containment controls or destruction techniques; and

3. worker exposure, particularly due to aerosols.

B. Program Structure

The OTS has expressed the need for methodology for assessing the accidental or deliberate release of genetically engineered microorganisms which could result in subsequent worker exposure and environmental contamination. The program proposed here addresses manufacturing processes based on genetically engineered microorganisms and subsequent use in other contained processes. Because genetically engineered microorganisms have already been developed for applications requiring deliberate release into the environment, the proposed program also addresses the development of procedures for assessing the engineering aspects of introduction of genetically engineered microorganisms into the environment for a number of such applications in the form of "scenarios" appropriate to the environmental conditions likely to be encountered at representative sites. Applications to be considered for evaluations include: agricultural formulations; pollutant clean-up/control (eg, spills, landfills, contaminated sediments, oil spills); tertiary oil recover; in-situ mineral recovery (metals leaching, oil shale), and other such operations not contained in chemical processing equipment in the traditional sense.

The engineering assessment protocols for release and exposure can be structured to account for several sets or combinations of various biological properties, or subsets, and appropriate applications involving deliberate environmental release, which the Workgroup and Review Committees feel should be addressed at this stage. Further, effort will be devoted to identifying those specific data (chemical, physical, and biological) which will be required as inputs to the engineering risk assessment protocols so that the OTS can specifically require the development and submission of such data as part of the PMN review procedures.

The proposed program is presented in detail in Section IV.D and is comprised of two major subdivisions:

(1) Subsection IV.D.1: Methodology for assessing the potential for accidental and deliberate release, exposure, containment and decontamination of genetically modified organisms from biologically based manufacturing processes, including the production site, shipment, intermediate storage and subsequent process use.

(2) Subsection IV.D.2: Methodology for assessing the potential for accidental and deliberate release from the site of application, worker exposure, containment and decontamination and for developing criteria for safe use in the deliberate introduction of genetically engineered organisms into a specific environmental area.

These items constitute the basic engineering elements of estimating the potential for accidental or deliberate discharge, and, in the event of process discharge, containment, worker protection, and decontamination. Note that biological control, i.e., control based upon biological characteristics incorporated by design into the genetically engineered microorganism, is considered as an activity integral to the development of the organism. While a desirable attribute, biological control is therefore viewed as a function more appropriate to the microorganism development process and is not considered an environmental engineering pollution control function.

C. Proposed Approach

The engineering assessment protocol modules for release and exposure for PMN review do not require the detailed data on the properties of the specific genetically engineered microorganisms as required for the health and environmental assessment modules. It is recognized that a substantial data base exists which deals with fermentation production technology, physical containment methods, decontamination and destruction techniques, and worker exposure and protection.

This data base includes past experience and practices with (1) chemical engineering biotechnology based processes (particularly fermentation, the most likely process to be used); (2) DOD, NASA, DHHS (particularly RAC), and health care industry technology for handling potentially hazardous substances and wastes; (3) industrial techniques developed for fault analysis, redundant systems, component reliability testing and failure analysis, and (4) equipment manufacturers. Further, the insurance underwriting industry may be a valuable source of information related to the frequency of equipment failures resulting in losses and to loss prevention technology. Based upon the examination of these sources, coupled with a modest amount of pilot scale research to develop data on potential release quantities and characteristics for specific operations, we are confident that the methodology can be developed for: estimating the nature and quantities of industrial process release, potential worker exposure, estimating quantities potentially released to the environment, evaluating containment and worker protection technology, decontamination technology, and monitoring strategies for environmental releases.

The project descriptions address the aspects of objectives, research approach, and project output in such a way as to respond to the OTS needs stated. Resource estimates are provided in the Program Summary provided for each major subdivision. It is estimated that 6 to 8 months will be required to work out special arrangements with other agencies for cooperation or participation, setting up Interagency Agreements where required, preparing scopes-of-work and obtaining OTS reviews. The normal lead time for acquiring contractor support varies from 6 to 8 months after the work is defined and authorized. However, this may be reduced through use of parallel efforts such as announcement of intent in the Commerce Business Daily, etc, and using existing Level-of-Effort contracts where appropriate. In general, the projects will take about 9-12 months of technical effort, 2 months for special arrangements for acquiring "sensitive" information, and 3-4 months for preparation and review of the draft final reports. Therefore, the projects will require about 18 to 24 months from initiation to completion of a report draft ready for publication.

IV.D.

DEVELOPMENT OF ENGINEERING INFORMATION AND METHODOLOGY FOR RISK ASSESSMENT, REDUCTION AND MANAGEMENT FOR GENETICALLY ENGINEERED MICROORGANISMS IN BIOLOGICALLY BASED MANUFACTURING PROCESSES AND DELIBERATE ENVIRONMENTAL RELEASE

PROGRAM AND RESOURCE SUMMARY

PROGRAM/PROJECT		RESOURCES REQUIRED		
	Positions (FTE)	S&E ($K)	Xmural ($K)	Total ($K)
IV.D.1 Biologically Based Manufacturing Processes				
.1 Potential for Industrial Process Release	0.8	48	625	673
.2 Potential for Worker Exposure	0.3	18	200	218
.3 Containment of Process Equipment Releases	1.0	60	250	310
.4 Identify Monitoring Needs and Stragegies	0.3	18	175	193
.5 Decontamination Technology	1.0	60	350	410
.6 Worker Protective Equipment	0.6	36	200	236
(1) Personal Protective Equipment	(0.3)	(18)	(100)	
(2) Personnel Isolation & Decontamination	(0.3)	(18)	(100)	
Totals	4.0	240	1,800	2,040
IV.D.2 Deliberate Environmental Release				
.1 Site Profile Evaluation Procedure	1.0	60	400	460
.2 Site Containment Alternatives	0.5	30	250	280
.3 Monitoring Needs and Strategies	0.5	30	125	155
.4 Site Decontamination Alternatives	0.5	30	200	230
.5 Evaluation oof Applications (Innoculation) Technologies	0.5	30	200	230
Totals	3.7	224	1575	1799

IV.D.1   Accidental and Deliberate Release from Biologically Based Manufacturing Processes

PROJECT: IV.D.1.1 Potential for Industrial Process Release

OBJECTIVE: Develop the procedures to predict the nature, source, and amount of accidental and deliberate release of genetically engineered microorganisms from industrial manufacturing plants manufacturing or utilizing such organisms.

APPROACH: The potential for industrial process release of genetically engineered microorganisms will be evaluated through a multi-task approach based primarily upon information available from the literature and experts augmented by limited experimentation as required to fill in gaps.

Task 1: Estimate the source and nature of emissions and effluents for deliberate process releases which would occur in the routine operation of a fermentation process. Another type of deliberate process release occurs when equipment is opened for charging of feed materials or maintenance. These latter items are to be covered in Task 2 as they are intimately related to the specific manufacturer's equipment design.

This task requires the analysis of the unit operations and processes and associated types of process and pollution control equipment which are likely to be used in the industrial biochemical engineering processes; the feedstocks, intermediates, by-products, and impurities and their chemical and physical properties; operating conditions; potential discharge points for deliberate

releases; types of release (hydrosol or aerosol) and expected
composition and properties. This task will be developed on
the basis of available knowledge of the chemical and physical
properties and process operating conditions determined in lab
and process development work with genetically engineered
organisms as a first choice. Where such information is
lacking, it will be developed on the basis of conventional
(i.e., non-genetically engineered organisms) biochemical process
information.

This task has been partially completed. A draft report
has been produced (Industrial Process Profile for Environmental
Use: Industrial Applications of Recombinant DNA Technology -
Jacobs Engineering/Technichron, 1983) on the unit operations
and processes, lab operations, and identification of chemicals
used in lab RDNA experiments. This draft report is the
starting point of a more detailed and explicit investigation
which will provide information relevant to identification of the
points and nature of deliberate releases of emissions and
effluents, quantities, and expected components, including those
relating to sampling for process monitoring and control and
product quality control.

Task 2: Estimate the potential for accidental or deliberate
release from processes and the characteristics of releases
for non-routine operations and equipment failures. For a
deliberate release, the specific equipment items must be
analyzed to estimate the potential release on the basis of

specific equipment design. In regard to accidental releases, for example, such as that occurring from a pump seal leak or a catastrophic event such as a pressure vessel or safety vent rupture, the most straight forward way to develop a "real" data base is to enlist the cooperation of specific existing plants (not necessarily using genetically engineered organisms) to furnish access to maintenance records. As a secondary approach, equipment failure analysis techniques might be applicable to determining the point of failure. Alternative approaches would be explored with the appropriate committees of professional and standards setting organizations.

Task 3: Estimates of the quantity of released material for deliberate and accidental releases can be derived in a fairly straight-forward manner through engineering analysis of the process unit operations and the manufacturer's equipment designs for both steady-state and start-up/shut-down conditions, including raw materials additions and maintenance.

```
RESOURCES:   Task 1, to complete -
                Intramural -  0.3 FTE, $ 18K S&E
                Extramural -           $150K R&D
                Total      -  0.3 FTE, $168K

             Task 2
                Intramural -  0.4 FTE, $ 24K S&E
                Extramural -           $350K R&D
                Total      -  0.4 FTE, $374K

             Task 3
                Intramural -  0.1 FTE, $  6K S&E
                Extramural -           $125K R&D
                Total      -  0.1,     $131K

                Grand Total   0.8 FTE, $673K
```

PROJECT: IV.D.1.2 Potential for Worker Exposure

OBJECTIVE: Develop the capability to estimate the nature of worker exposure from the deliberate and accidental release of genetically engineered microorganisms from industrial manufacturing plants.

APPROACH: The potential for worker exposure within the plant area can be estimated by an analysis of the plant operational procedures and related worker activities, eg., data logging, equipment operation, materials charging, quality control sampling, filter changing, and maintenance/repair activities. The analysis will address the frequency, duration, and route of potential exposure with estimates of the levels of maximum and average exposure doses (assumes Project 1, Task 3 is completed). The analysis will identify chemical, physical, and other properties of the materials processed which will be needed in the PMN review process to evaluate the applicabiity of specific alternatives.

This project will be coordinated with the National Institute for Occupational Safety and Health (NIOSH). The NIOSH (DHSS/USPHS/CDC) will be requested to join in this research project because of: (1) their mission to develop criteria for worker exposure as the basis for OSHA standards, (2) their experience and expertise, and (3) their ability to acquire worker and industry participation through the tripartite committee approach. Models for the extent of environmental contamination and general population exposure (transport, fate, and receptor relationships) will be required to carry this aspect of the risk analysis further. However, this type of activity more appropriately is handled by OEPER, based on the OEET plant emission/effluent estimates.

RESOURCES: Intramural - 0.3 FTE, $ 18K S&E
            Extramural - $200K R&D
            Total   - 0.3 FTE, $218K

PROJECT:	IV.D.1.3 Technology for Containment (or Engineering Controls) of Process Equipment Releases
OBJECTIVE:	Evaluate the potential performance and costs of alternative containment technologies for preventing/reducing the amount of accidentally or deliberately released genetically engineered microorganisms which would result in worker exposure or entry into the environment.
APPROACH:	These evaluations will be based upon engineering analyses of biochemical engineering processes, storage and shipment activities to determine alternative technologies for containment of deliberate and accidental releases, including ventilation air, spills clean-up, and engineering controls for worker protection. Estimates of the costs of containment methods and analysis of the impacts on process operability will be prepared to aid those responsible for risk estimation and risk reduction cost-benefit analysis in the identification of economically viable alternatives. A broad range of technologies will be evaluated including conventional chemical engineering and pollution control techniques and also those techniques employed in applications requiring extremely stringent controls and safety measures, specifically the military CBW (chemical and biological warfare) and other high-hazard biological activities. Selection of equipment based on optimization of choice to

minimize the accidental and deliberate release and opportunities for process modifications are additional alternatives which will be addressed. For example, pumps driven through magnetic impellers would eliminate the problem of leakage through the shaft seal or, alternatively, positive shaft seals or microfilters can be employed. The analysis will be used to identify chemical, physical, and other properties of the material processed as required in the PMN reviews to evaluate the integrity of materials of construction and effectiveness of the containment systems. We would invite NIOSH participation in the area of engineering controls for worker protection. The NIOSH engineering control group is located in Cincinnati, is currently working in this area, and has joined us in sponsoring other studies (e.g. Industrial Process Profiles for Environmental Use: RDNA; EPA Draft Report, 1982).

RESOURCES: Intramural - 1.0 FTE, $ 60K S&E
                Extramural - $250K R&D
                Total   - 1.0 FTE, $310K

PROJECT: IV.D.1.4 Identify Monitoring Needs and Strategies

OBJECTIVE: Identify those plant, process, and worker aspects which represent a known (deliberate discharge) or potential (accidential discharge) interface with the environment through which a pollutant of concern might escape into the general environment.

APPROACH: Operational analysis of the plant and worker activity patterns as established in those tasks addressing (1) potential release and worker exposure and (2) containment and decontamination will be used to predict environmental interfaces of concern. A monitoring strategy for detection of the pollutants of concern, or a suitable surrogate which can be used as an indicator of the potential for discharge, will be developed. A strategy for employing simulants to test the system periodically to ensure integrity will be included. This strategy can be combined with the measurement methodologies developed by OMSQA to serve as a guideline for PMN reviewers. The OMSQA and other knowledgeable groups will be requested to provide for Peer review of the strategies formulated.

RESOURCES: Intramural - 0.3 FTE, $ 18K S&E
Extramural - , $175K R&D
Total   - 0.3 FTE, $193K

PROJECT: IV.D.1.5 Decontamination Technology

OBJECTIVE: Evaluate alternative decontamination technologies which could be used in the destruction of genetically engineered microorganisms and hazardous/toxic chemical by-products from the production/use process for probable effectiveness and environmental acceptability.

APPROACH: Alternative technologies for decontamination (destruction of genetically engineered microorganisms and chemical detoxification, where required) of the substances recovered from various containment alternatives after deliberate or accidental process releases will be evaluated for effectiveness, estimated costs for installation and operation, secondary pollution and environmental acceptability. Methods developed for use in the conventional biochemical processing industry, NASA, medical disinfection, and DOD for biological decontamination will be evaluated in conjunction with environmental pollution control requirements, particularly hazardous waste disposal requirements and predisposal treatment options. Some alternative technologies include: thermal treatment (e.g., incineration, autoclaving), chemical (acid/base), bacterial degradation (aerobic/anaerobic destruction), various types of radiation, ozone, as well as others.

RESOURCES: Intramural - 1.0 FTE, $60K, S&E
            Extramural -   350K, R&D
            Total   - 1.0 FTE, $410K

PROJECT: IV.D.1.6 Worker Personal Protective Equipment

OBJECTIVE: Evaluate the probable effectiveness of available worker personal protective equipment to provide adequate protection in the event of an accidental or deliberate release of genetically engineered microorganisms, including accompanying chemical substances and contaminants, and biotic or abiotic factors involved in alternative decontamination procedures.

APPROACH: This project will combine information from a number of sources to evaluate the adequacy of worker personal protective equipment and technology for resistance to chemical and biological agents under expected contaminant release conditions. Protection against chemicals, particularly new chemical substances subject to PMN review are being examined under an ongoing project for OTS. Also, the NIOSH, EPA, and industry have conducted studies and developed standards which will serve as a basis for chemical exposure evaluation. The NASA, DOD, USPHS, and industry technology for protection against pathogenic contamination will serve as a valuable basis for biological exposure protection evaluation. It will be necessary to evaluate the technology from the aspect of protection from both chemical and biological agents under the conditions expected to prevail in various industrial exposure scenarios. In addition, alternative management techniques for containment and decontamination in the event of exposure will be aspects and limitations of the alternatives based on limitations imposed by chemical or biological agent properties. These will serve to determine what chemical, physical, and other properties of the genetically engineered microorganism and attendant substances must be furnished for the PMN reviewer.

RESOURCES: Task 1 - Personal Protective Equipment

    Intramural - 0.3 FTE, $18K, S&E
    Extramural -   100K, R&D
    Total   - 0.3 FTE, $118K

Task 2 - Personnel Isolation & Decontamination

    Intramural - 0.3 FTE, $ 18K S&E
    Extramural - $100K R&D
    Total   - 0.3 FTE, $118K

Grand Total - 0.6 FTE, $236K
            plus NIOSH contribution

IV.D.2   Deliberate Release into the Environment

PROJECT: IV.D.2.1   Site Profile Evaluation Procedure

OBJECTIVE: Identify any characteristics of the site that would influence proper containment and decontamination (if required) of the genetically engineered microorganisms, its genome, or the products of its metabolic processes.

APPROACH: In an approach similar to that employed in the preparation of an Environmental Impact Statement, an evaluation methodology and checklist of items to be considered in an evaluation of the potential for escape of the genetically engineered microorganisms, genome, or by-products from the field site will be developed. The methodology will address the potential for off-site migration in the various media during application, biological activity period, and decontamination activities through analysis of the site's environmental characteristics and identification of the various routes of potential migration (surface runoff, access to ground water, soil porosity and permeability rate, wind velocities, wildlife vectors, etc.) and specify specific migration routes of environmental concern which would require mitigation. In the event that a breach in the containment security system should occur which could permit an accidental or deliberate release from the site, remedial measures will be required. The issues of rapid deployment of emergency containment and decontamination techniques (as identified under tasks IV.D.2.2 and .4) will be addressed.

```
RESOURCES:   Intramural  -  1.0 FTE, $ 60K S&E
             Extramural  -         , $400K R&D
             Total       -  1.0 FTE, $460K
```

PROJECT: IV.D.2.2 Site Containment Alternatives

OBJECTIVES: Determine suitable site containment technologies to prevent uncontrolled migration of the genetically engineered microorganisms from the treated site boundaries after application.

APPROACH: Based on the migration routes and environmental concerns which are identified in project IV.D.2.1, alternative technologies and operational procedures will be identified and evaluated for potential cost and effectiveness for preventing migration during the genetically engineered microorganism activity and decontamination stages. The susceptability of the various engineering materials used for containment to environmental degradation, especially in regard to the genetically engineered microorganisms, decontamination agents, and native biota will be addressed also. Technology developed in the DoD CBW field test programs, the Superfund Remedial Action program, and the Emergency Response Program will be primary information sources.

RESOURCES: Intramural - 0.5 FTE, $ 30K S&E
Extramural - , $250K R&D
Total - 0.5 FTE, $280K

PROJECT: IV.D.2.3 Monitoring Needs and Strategies

OBJECTIVE: Develop requirements for monitoring to ensure site containment integrity, or, in the event of an uncontrolled release, trigger contingency plans.

APPROACH: This project will identify requirements for the development and employment of monitoring procedures to ensure that the site is secure in regard to pollutant migration via worker, environmental, or wildlife transport. The requirements will be identified through the analysis of the outputs from Tasks IV.D.2.1, IV.D.2.2 and IV.D.4.1 and existing environmental monitoring technology. A generic protocol will be developed that addresses informational needs to be satisfied, including those related to environmental monitoring, in conjunction with the OMSQA. The OMSQA and other knowledgeable groups will be requested to provide for peer reviews of the strategies formulated.

RESOURCES: Intramural - 0.5 FTE, $ 30K S&E
            Extramural -   , $125K R&D
            Total   - 0.5 FTE, $155K

PROJECT: IV.D.2.4 Alternative Site Decontamination Techniques and Procedures

OBJECTIVE: Determine suitability of alternative techniques and applications procedures for decontamination of the site.

APPROACH: Identify and evaluate suitable techniques for decontamination of the field site upon completion of the genetically engineered microorganism activity stage. This task will draw upon the output of Task IV.D.1.5, Decontamination Technology, and, in addition, explore techniques which are specifically designed for field applications (e.g., time release biocides). Procedures for application to specific scenarios based on various environmental conditions of concern identified in Task IV.D.2.1 will be evaluated for effectiveness in decontamination, prevention of secondary pollution, safety, and cost.

Prior to conducting a field application, it is important to determine what abiotic and biotic factors are effective in controlling or inactivating genetically engineered microorganisms that have been found to pose a potential medical or ecological threat. Abiotic factors would include: chemical disinfectants such as chlorine and other halogens, chlorine dioxide, ozone, other oxidizing agents; physical disinfectants such as ultraviolet light, electron beam radiation, ultrasonics, etc.; field conditions, such as pH, temperature, moisture content, etc.; barriers. Biotic factors include effect of predators, parasites, antagonists and competitors. It is also important to assess the effects of nutrient availability and concentration on the survival and persistance of the genetically engineered microorganisms in the proposed ecosystem.

The choice of the decontamination procedure depends on the type of environment in which the field test will be conducted. For example, for a closed environment, such as a wastewater treatment plant that has been seeded with the genetically engineered microorganisms, the chemical and physical disinfectants may be the appropriate and likely choice. In an open environment, however, such as a hazardous waste site, decontamination is more difficult. Reliance on field conditions, barriers, biotic factors, nutrient availability, and time in such an open environment may furnish an effective method of preventing, controlling and/or the multiplication and persistence of the genetically engineered microorganisms. The potential products, including decomposition products, resulting from decontamination require evaluation to ensure that hazardous wastes either do not result or are properly managed. Techniques which could be rapidly deployed in the event of a release will be identified. Also a list of information needs regarding specific properties of the gentically engineered microorganism and the formulation in which it is applied will be generated to determine physical, chemical, and biological property data required in the PMN application for the reviewer (as necessary).

RESOURCES: Intramural - 0.5 FTE, $ 30K S&E
Extramural - , $200K R&D
Total - 0.5 FTE, $230K

PROJECT: IV.D.2.5 Application Technology for Genetically Engineered Microorganisms

OBJECTIVE: Evaluate the use of alternative equipment and techniques to identify application procedures and conditions for introduction (or innoculation) into the environmental site that will be safe for the operator and would prohibit escape of the genetically engineered microorganisms from the site during application.

APPROACH: Identify various application techniques and equipment and then evaluate for safety. Examples of approaches to be considered are: (1) in spray application, the relation of droplet size and wind velocity are important to prevent wind drift; (2) the suitability of encapsulated genetically engineered microorganisms or genetically engineered microorganisms incorporated into an immobile substrate which could be deposited on the ground surface, subsequently releasing the bioactive substance upon exposure to moisture.

RESOURCES:
Intramural - 0.5 FTE, $ 30K S&E
Extramural - , $200K R&D
Total - 0.5 FTE, $230K

MONITORING TECHNIQUES FOR GENETICALLY ENGINEERED MICROORGANISMS

David Glaser,[*] Tim Keith, Peg Riley, Geoff Chambers, John Manning,
Susan Hattingh and Ralph Evans

*Harvard University Museum of Comparative Zoology, Cambridge, MA
May 30, 1985

I. . INTRODUCTION

The first field tests of genetically engineered microorganisms (GEM's) are being scheduled for spring, 1985 (Bioscience News Update 1985). With the approaching release of GEMs into the environment, monitoring techniques will have to be established to follow 1) the environmental fate of the GEM, 2) the continued presence or absence of the recombinant DNA (rDNA) in the GEM, 3) the continued functioning of the rDNA (that is, continued gene expression), and 4) the transfer of the rDNA to new hosts. This information is required before the appropriate regulatory committee can establish the potential risks resulting from widespread release of the GEM.

It is the purpose of this document to discuss the design of monitoring protocols and to suggest research needs for their development. Classical microbiological and modern molecular techniques that will permit analysis of GEM and rDNA fates in the environment are available. Standard microbiological techniques will allow monitoring of GEM's and active rDNA, if there is a suitable assay for the gene product, for example an enzyme assay. However, these techniques will provide little information concerning the rDNA sequences specifically. Molecular techniques to assay the DNA itself will provide information about the presence of the rDNA, changes in the rDNA sequence and its location in the genome, the functioning of the rDNA gene, and its transfer to new hosts. Both molecular and microbiological techniques will be described in the following sections with respect to their potential value, efficiency, sensitivity, and relative cost.

A section on microcosm experiments has been added to this document to tie together the various monitoring procedures. Typical host-vector systems can be analysed in microcosm studies to test the accuracy of the proposed monitoring protocols. These systems can then be used as standards which the industries can follow when designing microcosm and field tests for their specific host-vector systems.

The specific monitoring techniques that will be most effective will depend, to a large degree, on the specific organism being monitored. Therefore, it is useful to briefly discuss the types of recombinant organisms which have been proposed for experimental release and commercialization. The different types of recombinant organisms are:

1) organisms with normal genes deleted; an example of this is *Pseudomonas syringae* and *Erwinia herbicola*. The latter, produced by mutagenesis, has already been field tested by Microlife Technics, Inc. These bacteria lack a functional cell-surface protein that in wild-type bacteria catalyzes ice-crystal formation (Genewatch 1983).

2) organisms with new genes inserted using plasmid vectors, including:

   a) plasmid-mediated gene transfer in bacteria; an example of this type of GEM is the strain of *Pseudomonas cepacia* that grows on 2,4,5-T as sole carbon source. This was created by plasmid-assisted molecular breeding, that is, by combining several strains of bacteria with different plasmids coding for degradative pathways for various chlorinated hydrocarbons and for resistance to various antibiotics, along with several strains from waste dumping sites. These were grown together in a continuous culture with 2,4,5-T and other plasmid substrates. After eight to ten months, organisms capable of using 2,4,5-T as sole carbon source were isolated (Kellogg et al. 1981).

   b) plasmid-mediated gene transfer to organisms other than bacteria; an example of this type of GEM is the bacterial genus *Agrobacterium* which can transfer some of its plasmid DNA to infected dicot plant cells. There are two species of *Agrobacterium* of industrial interest, *A. tumefaciens* which carries the tumor inducing (Ti) plasmid and *A. rhizogenes* which carries the the root inducing (Ri) plasmid. Portions of both Ti and Ri plasmids can become stably integrated into dicot plant chromosomal DNA. These plasmids have been modified so their disease-causing genes have been activated thereby rendering them very useful vectors. There have been several successful introductions and expressions of bacterial genes in plants (Chilton et al. 1983, Herrera-Estrella et al. 1983, Schell et al. 1983). In one case the cells have regenerated whole plants in which the foreign genes were still expressed (Fraley et al. 1983). Kemp (1983) reported the first transfer and expression of a plant gene from one plant species to another; the seed storage protein from beans was successfully introduced into both sunflower and tobacco cells.

3) transposition-mediated gene transfer; transposable elements are DNA sequences that are able to replicate and insert copies of themselves at new locations in the genome. Experiments using transposable elements have been done with corn cells (Peacock, 1983).

The following sections will detail currently available monitoring approaches. It should be emphasized that there is no one approach to monitoring that will be suitable for all host-vector systems. Protocols will have to be developed that are suitable for following the fate of each GEM produced.

II. SAMPLING CONSIDERATIONS

A. *Introduction*

Developing an optimal sampling strategy is not trivial, because it depends on the organism, the gene product of interest and the medium from which it is to be sampled. Thus, monitoring strategies must be tailored to each specific case. However, there are considerations common to all monitoring problems. Certain scientific issues can be explored with test organisms to provide a basis for conducting specific monitoring assessments most effectively; these are discussed below.

B. *Qualitative sampling*

Two types of trade-offs exist in sampling strategy: (1) qualitative *vs.* quantitative, and (2) extensive *vs.* intensive sampling. (1) The term "qualitative" is used to describe sampling for presence/absence only. "Quantitative" sampling refers to attempts to enumerate organisms, plasmids or DNA sequences. (2) The phrase "sampling extensively" refers to sampling a large geographic area, while "sampling intensively" is used to refer to searching only a small area, but attempting to cover as many different habitats as possible, using closely-spaced samples.

Experience demonstrates that it is possible to enrich for most microbial activities in most environments. In large measure this is attributed to the fact that microbes disperse well and can last a long time under poor conditions. Furthermore, under favorable conditions, few organisms can quickly give rise to a large population. Beijerinck summarized this in the principle of microbial ubiquity: "Everything is everywhere; the environment selects." (Atlas and Bartha 1981, p. 5). Therefore, for the prediction of potential impact, it is of primary importance to find out where the GEM and rDNA sequence are transported and to determine how long they can last after their purpose is fulfilled, that is, to sample qualitatively and extensively. This is in contrast to the needs of the producers and users of GEM's, to whom it is very important to know the density and activity level of the GEM in the target field, that is to sample quantitatively and intensively.

This suggests that research should concentrate on methods that are good at very low densities to allow qualitative and extensive assays for the presence of small populations.

C. Desorption from sediments

To sample from sediments and soils one must deal with the problem that the organisms are attached to surfaces. A sample of the community can be transferred to fluid phase simply by placing a bit of soil in fluid, shaking vigorously, and then spreading a sample of the fluid on an agar medium for plate counts. However, organisms of interest may not disassociate readily from particles. Thus, it is necessary to maximize the ability to desorb cells from surfaces. There has been much work on methods of desorption (see, for example, Marshall 1976). The first research need is for compilation of existing methods. Next, experimental research into whether desorption actually increases plate counts is needed. If it is useful, it will then be necessary to tailor methods to the types of organisms likely to be released.

D. Enrichment

An alternative to methods for plate counting low-density populations may turn out to be enrichment. Standard enrichment techniques aim to select for the organism of interest. For some GEM's, however, this requirement need not be so strict. For example, if an organism produces an unusual organic molecule, enriching for the organism may increase the concentration of the product, and this might occur independently of any increases in other populations. Enrichment methods can be quantified with Most Probable Number (MPN) methods (Russek and Colwell 1983), although large samples are needed to achieve reasonable precision.

Enrichment is sometimes more effective as a monitoring technique than plate counting. For example, Kilbane et al. (1983) added a strain of bacteria that could grow on 2,4,5-T as sole carbon source to a 2,4,5-T-contaminated soil and watched the strain decline in numbers as the substance was catabolized. They could not detect any organisms on plates after 8 weeks. At 12 weeks, 2,4,5-T was again added to the soil, and by 15 weeks the 2,4,5-T-degrading strain became detectable again and increased dramatically in density. That is, the potential for 2,4,5-T degradation and population growth remained, even though plate counts were zero.

The problem of desorption of cells from particles in aquatic sediments and soils is probably not as great using broth enrichment techniques, because the sediment itself can be placed into the medium. Thus, two strategies present themselves for sampling from solid phases: (1) dissassociation of cells from particles by physical or chemical means, followed by selective plating; and (2) replicated enrichments, followed by replicated isolations from each enrichment flask. Organisms isolated in this manner can then be studied as necessary using molecular techniques.

A great advantage of the second method is that enrichment is likely to pick up GEM's at lower densities than is possible with direct plate counts. This suggests that the high densities present immediately following inoculation might be sampled with plate counts. If numbers then decline to levels undetectable with plate counts, broth enrichments can be used, quantified with MPN methods. Molecular analyses can be performed on both types of samples.

E. Partitioning in the environment

One objective of low-density sampling is to determine how added microorganisms partition themselves in the environment. When an organism is introduced to the environment, it tends to survive better in certain habitats or microhabitats than in others. For example, such habitats in a soil might include leaf surface, rhizosphere, the top centimeter of soil, anerobic microsites, invertebrate guts and external surfaces of invertebrates. *Pseudomonas fluorescens* and *P. cepacia*, strains of which have been engineered for release into the environment (see section VI), can act both as animal and plant pathogens (Bergan 1981). Thus, sampling should include potential local animal hosts.

In non-target locations only small numbers of GEM's will probably exist. This suggests that it may be worthwhile to sample initially very intensively and frequently from the target field and a small adjacent area, just following release of the organism. For example, if it is found that *Pseudomonas fluorescens* with a gene for a *Bacillus thuringensis* endotoxin tends to survive best in the top centimeter of grass rhizosphere, then sampling effort in other locales should concentrate (not exclusively) on the top centimeter of grass rhizosphere, because this is most likely to give positive results at low densities. The initial intensive sampling should seek to answer the following questions: Does the organism tend to move towards or survive better in certain types of habitats? What is the rate of increase in numbers in adjacent areas? The answers to these questions can be used to suggest where sampling effort might be most effective in non-target fields with very low densities. Research might be directed towards studying the usefulness of this type of protocol in test organisms.

F. Issues in sampling methods

Sampling methods have been studied for a long time, and many methods exist and are well-documented (for example see Atlas and Bartha 1981, Lynch and Poole 1979, Sieburth 1979). Effective sampling usually requires concentrating the organisms, for example by filtration through an 0.2 micron filter. Coliforms are routinely sampled by filtering a standard quantity of fluid through a filter, and then placing the filter directly onto M-Endo agar, which selects for Gram negative species and turns metallic green when a rapid lactose fermenter such as a coliform is present.

With respect to following the movement and survival of DNA sequences, "habitats" include other organisms, potentially both animals and plants. Thus, an important question is: How extensive should the search be for determining the transfer of a certain DNA sequence? It is possible for genes to be transferred from bacteria to plants, for example from *Agrobacterium tumefaciens* to many species of plants (Atlas and Bartha 1981). In addition, it is possible for plant pathogenic species to also be animal pathogens, for example *Pseudomonas fluoresens* and *P. cepacia* (Bergan 1981). A useful rule of thumb would be to tailor each protocol to each individual case. These are issues that are not amenable to quantitative cost-benefit analyses. The cost of such sampling must be balanced against the probability of such transfer occurring and the possibility and likely magnitude of adverse effects.

It may be worthwhile to keep the following in mind: (1) An important problem in quantitative sampling is deciding on the size of the sample. A good tool for designing an economical but sufficient strategy is the performance curve, which is a graph of density against sampling effort (for example numbers of samples). This curve might rise and fall as sample size rises and then level off at some larger sample size. To determine the best compromise between expense and accuracy, one should perform a greenhouse test, taking a large number of samples. As each sample is taken in sequence, the estimated density is plotted against the number of samples taken so far. The optimum sample size is the minimum number of samples required to reach the flat part of the curve.

(2) For some questions, relative numbers are just as useful as absolute densities and may be easier to measure with a desired level of precision. For example, the rate of die-off of an added population can be estimated with relative densities.

III. MONITORING TECHNIQUES

A. Conventional microbiological techniques

Conventional microbiology offers many methods for the isolation of microorganisms from mixed culture samples. Many of these methods are for the characterization of the total microbial population, and as such are of little use in the recovery of a specific genetically engineered organism that has been released to the environment. This section of the report will examine selective isolation of organisms on solid media and in liquid culture.

Solid media techniques involve the immobilization of the desired liquid enrichment media with agar. Mixed cultures are appropriately diluted in sterile media, and then "streaked" onto the agar plate containing the indicated media. The individual bacteria grow into colonies on the plate, which allows the analyst to pick the colony from the plate for further purification and identification. This technique is used by aquatic and soil microbiologists to identify species capable of growing on the chosen medium. Fermentation and industrial microbiologists use this technique to identify microorganisms capable of growing on the medium or producing a valuable byproduct from it.

The species recovered are dictated by the medium chosen, that is, its components and their concentrations. Therefore, the nutritional requirements and unique metabolic functions of the GEM should be well characterized.

To recover a GEM selectively on an agar medium, one can insert DNA sequences that code for unusual metabolic traits, such as: (1) ability to grow on an unusual compound; (2) resistance to a heavy metal, antibiotic or other compound; and (3) production of a specific compound that can be assayed with color changes in the medium. These markers are considered in section IIIC.

As discussed in section II, liquid broth enrichment techniques are probably more effective at recovering GEM's at low population densities and therefore are likely to be of great importance. These can be quantified with most probable number statistics (Russek and Colwell 1983), although large sample sizes are needed for reasonable precision.

B. **Immunological techniques**

   1. Standard Methods of Antibody Production

Fluorescent antibody techniques have been used for many years in medical microbiology and pathology to examine tissue samples for infectious agents. Recently, they have been applied in mixed culture environments to explore ecological questions. The uses and limitations of immunofluorescence techniques have been discussed by Bohlool and Schmidt (1980).

To develop a fluorescent antibody a pure culture of organisms (the antigens) are injected into a rabbit that then develops antibodies to the organism. The animal is periodically bled to isolate antiserum to the injected microorganism. After purification, the antibody is attached to a fluorescent dye, commonly fluorescein isothiocyanate (FITC). The labeled antiserum is then added to a sample from the environment, which might be a smalll quantity of soil, or a filter through which water has been passed. The antibody binds to the antigenic microorganism. In theory, when the preparation is viewed under epi fluorescence microscopy, only the antigenic organism with the attached antibody can be seen.

Immunofluorescence techniques have been used to follow strains of *Rhizobium species*, *Azotobacter*, *Beijerinckia*, *Azospirillum*, nitrifiers, sulfur- and iron-oxidizers, and various other types of organisms. Attempts have also been made to look at fungi (Bohlool and Schmidt 1980).

However, there are several problems with the techniques: (1) Bohlool and

Schmidt (1980) calculated that minimum countable densities in soils are about 10^6 to 10^7 organisms/g soil. This is far too insensitive for studying low population sizes of GEM's. The severity of the problem can be reduced by increasing the efficiency by which cells are desorbed from the solid phase. Therefore, research is needed on desorption techniques, as mentioned in section II for increasing the efficiency of plate counts.

In addition, the efficiency of recovery can be estimated in experiments in sterile soils. This allows calibration against simple non-selective methods, such as plate counts and epifluorescence microscopy with cells stained with dyes such as acridine orange (Hobbie et al. 1977) and 4'6-diamindino-2-phenylindole (DAPI; Porter and Feig 1980).

(2) The GEM cells may not be the only fluorescing particles in the environment. There are several reasons for this: (2a) Inorganic particles sometimes bind nonspecifically to the fluorescent antibody complex allowing soil particles to interfere. (2b) Material in the environment can autofluoresce. (2c) Other species may have surface proteins similar to those of the organism of interest. There are also limitations in the method of production of antibodies. The organisms injected into the rabbit may contain low levels of contaminant proteins. The isolation procedure may not be perfect, resulting in a mixture of different antibodies being added to the environmental sample. Specificity can be increased significantly by using monoclonally produced antibodies, which are discussed in the next section. There is an important need for research into the specificity of fluorescent antibody techniques in mixed cultures containing closely related species.

(3) On the other hand, GEM cells may not fluoresce. (3a) Organic slimes may prevent the immune reaction between the bacteria and the fluorescent antibodies. For example, while attempting to determine the number of nitrifying

bacteria in a fixed-bed wastewater treatment reactor, Szwerinski et al. (1985) found large numbers of fluorescent cells in enrichment cultures taken from the reactor. However, very few cells were seen in direct observations of environmental samples. MPN counts of enrichment cultures gave much higher counts than direct observations. This is likely to be a problem in any environment with organic slimes, most importantly wastewater treatment facilities. (3b) The antigen can become unstable under some growth conditions. Bohlool and Schmidt (1980) site several examples suggesting that this may not be a serious problem. However, stability should be tested for each antibody that is used.

(4) Finally, immunofluorescence may not tell us what we want to know. (4a) Live and dead cells cannot be distinguished. This might be seen as an advantage. Very low populations of live cells might be missed by these techniques; having dead cells present increases the probability of finding small populations. (4b) If the antigen is not a product of the rDNA, then this technique will not be able to tell if the original antigenic cells have lost the rDNA or if other strains have picked it up. Using two antibodies in the same GEM, one to a normal cell protein and one to an rDNA product, can potentially give information on rDNA transfer and loss.

In summary, fluorescent antibodies can potentially be extremely useful for monitoring GEM's in the field. However, several problems exist in sensitivity and specificity. As discussed in the next section, monoclonal antibodies can potentially alleviate much of the specificity problem, as well as reducing production costs considerably.

2. Monoclonal Methods of Antibody Production

Monoclonal antibodies (MAbs) can provide a very reliable and sensitive method for identifying and monitoring the gene products of GEM's. MAbs are prepared by injecting a mouse with a purified antigen, removing its spleen after several weeks, isolating the antibody producing cells (the B lymphocytes), and fusing them with mouse myeloma (tumor) cells. The products of this fusion are grown in a selective medium which allows only the hybrid cells (hybridomas) to grow. These cells are able to grow permanently in cell culture due to the tumor cells and produce a specific antibody due to the contribution of the B lymphocytes. Hybridoma clones that express the particular antibody of interest can then be grown either in vitro or in the abdominal cavity fluid of mice for the production of large quantities of monoclonal antibodies.

This method allows the production of large quantities of specific antibodies against given antigens. Unlike conventional immunological methods, antibodies produced by the monoclonal techniques are homogeneous and are therefore more reliable (fewer false positives) in detecting a specific antigen. However, as with conventional immunological methods, disadvantages include the inability to tell live from dead cells, autofluoresence of environmental material and potential interference from sources such as bacterial slimes in the environment.

MAbs can therefore be extremely useful in detecting the products of genes introduced or altered by rDNA technoloy. The gene product of the GEM is isolated, purified and used as the antigen to elicit an immune response and the production of antibodies against it. Providing that the gene product is a strong enough antigen to provoke an immune response, the use of MAbs will allow a sensitive method for detecting its presence.

Currently the Damon Biotech Corporation is developing a new technique for the production of large scale amounts of MAbs by the technique of microencapsulation. The method uses a porous carbohydrate capsule to surround the hybridoma cells thereby retaining the antibodies produced while allowing the circulation of nutrients and metabolic wastes. When the encapsulated colonies are harvested after several days, the growth medium, according to Damon, contains up to 40-50% by weight of MAbs. This new production method should greatly reduce the cost of MAbs and make the use of MAbs an economic way to detect the presence of a particular gene product.

The use of monoclonal antibodies should be considered a very sensitive, reliable, and cost efficient means to identify and monitor the gene products of GEM's providing that antibodies to the gene product can be elicited through an immune response.

C. The use of genetic markers

The use of marker genes may prove to be a very sensitive and inexpensive way to follow GEM's upon release into the environment. There are several types of marker genes available to genetic engineers:

(1) chromogenic markers; these are genes that produce a precursor to a biochemical pigment. Using appropriate media one can produce pigmented microorganisms. This confers a scoreable phenotype on the GEM (Masui et al. 1984).

(2) resistance markers; these are genes that provide some type of resistance to the microorganism, eg. antibiotic or heavy metal resistance. These markers confer a selectable phenotype to the GEM; such genes may be favored in some natural environments and their spread could have unexpected consequences.

(3) rare sugar or other rare carbon sources; these are genes that confer the ability to use unusual sugars or other carbon sources. One example would be lactose utilization. Few naturally occuring bacteria other than *Escherichia coli* can ferment lactose, making lactose fermentation a marker which would differentiate a recombinant strain from most naturally occuring strains.

Markers used to label GEM's can be placed on the chromosome or on plasmid DNA. One would want to link the marker gene and the rDNA gene to limit the possibility that the marker is separated from the rDNA gene by recombination. A marker situated on a plasmid will generally tend to be less stable for several reasons; 1) conjugative plasmids may be transferred infectiously between the GEM and other species in the environment, 2) if non-conjugative, the plasmid may be mobilized by mobilizing plasmids that may be acquired by the host strain from other species, and 3) if the plasmid contains transposons, these elements may lead to the transfer of genetic information between species. Some initial studies have examined plasmid transfer in some commonly used host-vector systems (Anderson, 1975; Sagik and Sorber, 1979; Levy *et al*., 1980; Schiff and Klingmuller, 1983; Levy, 1984).

Marker genes can be constructed with various degrees of specificity. As mentioned previously, lactose fermentation may be an efficient marker gene to distinguish between a GEM and naturally occurring bacteria (other than *E. coli*). Lactose fermenting microbes can be identified very simply by using the coliform counting technique. This involves plating out samples from nature on M-Endo agar. This agar contains lactose and dyes to inhibit gram positive bacteria. The agar turns metallic green when the medium turns strongly acidic, as is the case when *E. coli* or any rapid lactose utilizer is plated on it.

When using any one metabolic characteristic for selective plating, one runs the risk of recovering naturally occurring organisms with the trait, as well as naturally occurring organisms to which the rDNA has been transfered from the GEM. To increase specificity, more than one marker can be inserted into a GEM, for example the ability to use an unusual compound and resistance to an unusual antibiotic. In addition, if the rDNA is on a plasmid, then having one marker on the plasmid and one on the chromosome allows one to trace separately the fates of the plasmid and the GEM.

To make a marker gene more specific it could be placed under the control of some promoter-operator sequence which would allow the gene to be induced under conditions unrelated to a need to ferment lactose. One example would be the trp promoter-operator region, which would allow the induction of the gene only when tryptophan is present. A second example would be the phage lambda's right operator, which would allow induction only in response to UV radiation or high temperature if a temperature sensitive repressor were used. Promoter-operator units such as these can be engineered so as to allow induction under known laboratory conditions for monitoring purposes, but to prevent induction under conditions leading to the selection for the marker in the field.

Genetic engineers have on hand a variety of marker genes that can be made as specific as desired. These marker systems can then be very easily tested in microcosm or greenhouse experiments and the efficiency and sensitivity of each marker system analyzed directly. This technique may prove to be an inexpensive first step in most monitoring protocols. It would then be necessary to examine the GEM's isolated in the first step for continued presence of the rDNA gene.

D. **Molecular techniques**

  1. Restriction enzyme mapping

The use of restriction enzymes to generate a restriction map of host DNA can be a very useful technique for identifying and monitoring recombinant organisms, especially those with genes introduced by plasmids. The key to restriction mapping lies in the ability of certain enzymes found in bacteria to cut double-stranded (ds) DNA at specific sites. Each restriction enzyme recognizes a specific nucleotide sequence between 4 to 6 base pairs in length. The enzyme cuts the DNA each place the specific sequence occurs. Different enzymes recognize and cut different nucleotide sequences. A piece of DNA, cut with a restriction enzyme, will produce a number of distinct fragments. These fragments can be separated on the basis of size by gel electrophoresis. The fragments can then be visualized by staining the gel with ethidium bromide (which binds to DNA) and photographing it under UV light (which causes the ethidium bromide to fluoresce). The size of the fragments can be determined by running a control on the gel which consists of DNA fragments of known size. By sequentially cutting the DNA with a series of restriction enzymes one can construct a map of specific restriction sites (Meyers et al., 1976).

For use in monitoring, one would first have to construct a restriction map for the nonrecombinant and recombinant plasmids. A comparison of the two maps would show where the novel gene had been inserted into the plasmid DNA. Once the restriction map of a plasmid with an inserted gene is known it should be fairly easy to monitor this recombinant plasmid. One would first have to isolate the plasmid DNA from the host bacteria sampled in the wild and then use the appropriate restriction enzymes to analyze it. The pattern of fragments produced can be compared to the original maps of both the normal and recombinant plasmids to verify the continued presence of the rDNA insert. Inserted and deleted DNA would be recognized as changes in the size of particular DNA fragments.

The method is sensitive to the detection of inserted and deleted genes and is fairly rapid. The restriction enzymes themeselves are expensive but if the manufacturer is required to do the initial analysis of the non-recombinant and recombinant plasmids, only a few enzymes should be necessary for subsequent plasmid characterization.

2. DNA probes

Classical microbiological techniques will allow sampling of the environment for novel organisms. It is then important to analyze the DNA of these organisms to verify that the rDNA is present, functioning, and in its proper location. In addition, tests may be required to examine the rDNA for transfer into new hosts (or new positions in the original host DNA). These analyses require specific nucleotide sequence analysis. This can be accomplished by using DNA probes. When the rDNA is being characterized and inserted into the host, a probe (DNA sequence specific to the rDNA) can be constructed for use in future monitoring events. A probe is essentially a complementary piece of DNA that will hybridize (base pair) to the rDNA when both pieces of DNA are mixed together in single-stranded form. The presence of this hybridization event can then be visualized in two ways; through the use of radioactive or biotin labelling. The former requires that the probe sequence incorporate radioactive nucleotides (usually ^{32}P) into its sequence. The presence of the probe can then be followed with the aide of x-ray film. The DNA of interest is attached in single-stranded form to a filter, the probe DNA (in single-stranded form and with attached radioactive nucleotides) can then be washed over the filter. A piece of film is then laid over the filter, and wherever hybridization occurred a spot will show up where the radioactive nucleotides exposed the film (Maniatis et al., 1982). This technique is expensive due to the cost of radioactive nucleotides and involves the handling of hazardous material. In addition, the probe is short-lived (the radioactivity decays fairly rapidly).

Biotin labelling of probes is currently being developed as an inexpensive, non-hazardous alternative to radioactive labelling. This technique involves the incorporation of biotin into the nucleotide sequence of the probe and takes advantage of biotin-avidine binding. First, avidin binds to the biotin. Next, an enzyme binds to the avidin and converts a colorless soluble substrate into an insoluble pigment. Spots of pigment on the filter correspond to places where probe and rDNA sequences have hybridized (Lewin, 1983; Langer et al., 1981). This technique will have the same specificity as the radiolabelling.

The exact nature of the probe (length and sequence) will depend on the host-vector (HV) system being employed. The length of the probe sequence will have to be calibrated to ensure against both random hybridizations with the host DNA and specific hybridizations with similar genes. This is a particular problem when the rDNA is merely an altered gene normally present in the host. It will be less of a problem if novel genes are being introduced. It may be appropriate to require the rDNA manufacturer to analyze the probe specificity and document the percentage of false positives/negatives.

The probe can be used for monitoring in two different experimental approaches. These approaches are the focus of the remainder of this section.

a. Restriction fragment hybridization

When characterizing the novel organism with fingerprinting (see previous section), the DNA fragments from the agarose gels can be transferred to filters and irreversibly bound. The DNA can be denatured (so that it is single-stranded) and a probe can be washed over the filter. The probe will base-pair with the rDNA and when visualized (see previous section) the size of the exposed fragment can be compared with the restriction map produced from the original novel organism. If the rDNA is in the same position on the chromosomal or plasmid DNA then the band that lights up will correspond in size to the restriction fragment produced when the gene was originally introduced into the host. This technique will provide evidence 1) whether the rDNA is in the original position, 2) whether the sequence has been altered, and 3) whether it has been deleted. The technique may not be sensitive to minor sequence changes (point mutations) or even gross changes (inversions). An additional set of tests must be performed to ensure that the rDNA is still functional (see monoclonal antibodies section). Restriction fragment hybridization is an expensive test since it involves not only the use of restriction enzymes but also the creation and labelling of a probe(s).

b. Colony hybridization

An alternative to fragment hybridization is the use of colony hybridization (Grunstein and Hogness, 1975). This procedure involves the plating out of sample cells (host cells) onto agar plates. The cells are lysed (cell walls broken) and a filter is placed over the plate so that the DNA from the host cells can be bound directly from the plate surface. The filter can then be probed in the same manner described above. Film is placed on the filter and spots on the film corresond to colonies on the original plate that contained DNA which hybridized to the probe.

This test allows one to scan many cells from a sample to pick out those with the rDNA incorporated. The cells producing spots on the film can then be further characterized to be sure 1) the cells are the intended host of the rDNA, 2) the rDNA is still in the same position (fingerprinting), 3) the gene is functioning.

This technique allows a rapid screening of many samples in a relatively inexpensive fashion. It has recently been shown to be a highly sensitive technique capable of detecting one colony in 10^6 colonies of a nonhomologous DNA background (Sayler et al. 1985). With appropriate amplifying media even lower densities may be detected. It is likely to be the best technique to assay directly for rDNA.

3. DNA-DNA hybridization

The technique of DNA-DNA hybrization entails the study of the temperature dependent kinetics of the association of single stranded DNA molecules to form double stranded structures (duplexes). The technique has been utilised mainly in systematics studies and is capable of generating highly reliable data. However it has limited resolving power since DNA sequences which are similar cannot be discriminated as being different and sequences which are very divergent will not reassociate.

It is unlikely that the technique of DNA-DNA hybridization will be useful for detecting recombinant DNA molecules in GEMs released into the environment. However, it may provide an estimate of the extent to which a particular sequence has been altered following the release of some particular organism.

A detailed review of this technique has recently been published (Sibley and Alquist, 1983). Using this method purified probe DNA from a reference source is sheared by sonication to a length of 500 base pairs. The probe is then denatured by heating and subsequently labelled with ^{125}I. Trace amounts of the labelled probe are mixed with genomic DNA and the temperature at which it forms duplex hybrids with whole genomic DNA from the organism under study is measured.

The difference in temperature between experiments with homologous and heterologous combinations of probe and sample can be related to the degree of sequence mismatch between the heterologous probe/sample pair. The best estimate for this genetic distance is that a difference of 1 degree C corresponds to 1% sequence divergence. The lower detection limit is around 0.5-1.0% divergence and the upper level of divergence that can be reliably estimated is around 30%.

The technique of DNA-DNA hybridization may be capable of identifying whether genes inserted into GEM's are still present. The main disadvantages inherent in this technique are the need for substantial (>200 ug) samples of pure whole genomic DNA, the expense of the radioactive probe, and the limited resolving power. The more direct method of hybridizing DNA probes to colonies or to restriction digests is more sensitive and efficient.

4. Genomic sequencing

A new method of providing DNA sequence data from specific regions of chromosomes direct from whole genomic DNA was recently reported (Church and Gilbert, 1984). This technique offers a rapid and sensitive method for screening released recombinant clones for DNA sequence stabilty.

The method is based on a combination of chemical sequencing of DNA (Maxam and Gilbert, 1980) and Southern blotting (Southern, 1975). Whole genomic DNA is first obtained from a culture of the organism of interest. The DNA is cleaved with a selected restriction enzyme which cuts the DNA within 100 bases of the sequence to be investigated. The resultant mixture of DNA fragments is divided into aliquots which are then subjected to the separate chemical modification and chain cleavage reactions used in conventional chemical sequencing of DNA. The products of the individual reactions are separated by electrophoresis and then transfered to nylon membranes. The DNA can be permanently bonded _in situ_ by UV irradiation and the bonded blots are stable.

The DNA sequence of interest is cloned into M13 phage and single stranded DNA isolated. Radioactive probes (150 base pairs long) complementary to the cloned sequences are then prepared and hybridized to the nylon screen blot. DNA fragments binding the probe are revealed by autoradiography and the DNA sequence read from the photographs obtained.

Thus the presence or absence of an introduced gene could be monitored by determing whether any of the genomic DNA bound the radioactive probe prepared from the originally introduced sequence. If present, the actual DNA sequence could be compared with that of the probe to determine its fidelity.

This technique is very expensive and very new. It is potentially useful in fine structure determinations, but less expensive, better characterized methods exist, for example Sanger sequencing (Sanger et al. 1977). Therefore, it is not likely to be of use in the foreseeable future for routine monitoring of GEM's in the field.

IV. MICROCOSM TESTS FOR MONITORING TECHNIQUES

A. Microcosm construction

A considerable amount of effort is needed to apply the techniques described in the previous section to environmental situations. In order to accomplish this, microcosm experiments are necessary to evaluate the specific monitoring protocols for test organisms. These must ultimately be tested in the field. There has been much work on microcosm technology for many different habitat types, and much of this has been summarized and referenced in a series of books and symposia (Gillett and Witt 1979, Witt and Gillett 1979, Geisy 1980, Pritchard and Bourquin 1984, Kinne 1977, Draggan and Van Voris 1979). In addition, there exist several large-scale in situ aquatic systems which can be used as intermediates between laboratory microcosms and the field (Grice and Reeve 1982). In Canada, the experimental lakes region provides opportunities for perturbing and observing natural systems experimentally (Watson 1980).

It might be argued that in view of the shortage of funds for the problem of assessing effects of genetically altered organisms, it would be best not to use these funds for research into the design of new microcosm systems, because many already exist. However, one important aspect of regulation is standardization, and no standardized set of microcosms exists for any environment. In addition, microcosms must form an integral part of fate and effect studies. Thus, it would be well-advised to conduct research into microcosm design as an integral part of an experimental approach to monitoring techniques, fates and effects.

The problem of standardization of microcosm assay systems is a very difficult one. The use of only one standardized system may be a desirable approach, but it is not a scientifically credible one because of the limited variability of any one experimental system compared with nature. This is a problem because the techniques described above can perform differently under different natural conditions. Because of these considerations, several types of construction should be used, for example the various designs described in Witt and Gillett (1979) and Gillett and Witt (1979). In addition, geophysical parameters such as soil water content and light and temperature regimes, should be varied.

Eventually, a monitoring protocol may include a standard series of microcosms of varying construction, run under a set of standard conditions. However, even this amount of standardization may be impossible, because the environments into which GEM's are likely to be released vary greatly both geographically and ecologically. Thus, appropriate microcosm design might have to be decided for each case.

Research into microcosm design will undoubtedly be useful for other issues in environmental protection as well, for example fates and effects of toxic chemicals (Kimball and Levin 1985). Such efforts are likely to involve long-term projects, because the design of microcosm systems for measuring important ecological variables requires knowing which variables are important and how to interpret changes in them. This requires much fundamental knowledge of ecological processes which we do not now have. Thus, support for basic studies of microcosm construction should be long-term and should involve multi-disciplinary teams so that physical, chemical and biological interactions can be explored. Because of its relevance to other ecological problems, it may be appropriate to support such research jointly with other programs in environmental protection.

B. **Microcosm methodology**

There are two types of organisms to use in microcosm tests: (1) GEM's that are likely to be released; and (2) organisms that already are added to field systems, whether they are GEM's or not. For example, a rhizobium with various markers might be used to test monitoring methods. The advantage of these organisms is that field trials can be used to test the realism of microcosm tests.

The organisms used in microcosm tests can be marked in different ways. Several strains of a test organism, differing only in specific markers engineered into them, can be used to test the relative sensitivity of recovery methods and stability of marking methods.

Different monitoring techniques should be tested together. Part of these comparative studies might include sterile and gnotobiotic (known species composition) systems. This will allow one to use simple methods such as non-selective plating and general fluorescent stains to calibrate more specific methods such as immunofluorescence.

C. *Sample protocol*

The usefulness of microcosm experiments lies in the ability to use sensitive, expensive techniques to assay for GEM's and rDNA in order to evaluate the various methods of recovery and to examine the fates of rDNA and GEM's in model systems with greater precision than is possible in the field. A sample protocol is offered below as an example of an approach used to monitor a GEM.

1.   a.   Assay for live cells, if possible with selective plate counts.
       b.   Assay for rDNA activity with immunofluorescence techniques.

2. If plate counts decline to unmeasurable levels, use enrichment techniques to continue sampling for live cells.

3. Simultaneously with 1. and 2., assay for the presence of rDNA with colony hybridization methods. Comparison with the other techniques will help answer the question: Is the rDNA gene still expressed?

4. If it becomes clear that rDNA sequences can be found, but that activity cannot be measured, then the reasons must be found for the apparent lack of gene expression. These include:

    a. The structural rDNA has been altered.
    b. The control region for gene expression has been damaged.
    c. The gene has moved to a site at which it is not expressed.
    d. The rDNA has moved to different strains in which it is not expressed.

    These reasons can be explored using the molecular techniques: Restriction enzyme maps will tell if the gene has moved or if parts were deleted. The use of labelled probes with maps will tell this more precisely. Finally, genomic sequencing can be used to detect fine-scale changes.

D. <u>Containment</u>

One problem with microcosm assays for monitoring, fate and effects studies is the containment of the GEM and rDNA. Preliminary monitoring and fate studies should be done in the laboratory under carefully controlled conditions, involving the standard procedures for working with genetically-engineered microorganisms. It is important to test both monitoring techniques and ecological effects at this stage, because the information gathered will be needed to design greenhouse tests.

There are two aspects to containment in greenhouse experiments: methods taken to protect against accidental release and a monitoring protocol to assay for accidental release. The stringency of greenhouse controls should be determined by a combination of information on the basic biology of the GEM and the function of its rDNA, along with results from laboratory tests of potential ecological effects. For example, preliminary toxicity tests can be performed in the laboratory using single species and small microcosms. If results are all negative, then larger-scale greenhouse tests can be performed with more complex microcosms. Since any experimental test of ecological effects is insufficient to prove the impossibility of adverse effects, laboratory results must be combined with other biological knowledge. Two examples of GEM's designed for release into the environment are: (1) a strain of Pseudomonas cepacia designed to degrade 2,4,5-T, a chlorinated hydrocarbon; and (2) a strain of Pseudomonas fluorescens with a gene for the Bacillus thuringensis endotoxin inserted, designed to poison a species of pest caterpillar. As discussed in section VI, general biological knowledge suggests that both the severity and likelihood of potential adverse effects of the second are greater. This suggests that controls in greenhouse tests should be tighter and that a monitoring protocol should be well-tested in the laboratory before greenhouse tests of the P. fluorescens are initiated.

It is necessary to be able to monitor for the escape of GEM's or rDNA from greenhouses during fate and effects tests. The results of laboratory tests of monitoring techniques should be used to design a greenhouse monitoring program. This interim monitoring program may not be the same as that used in field tests or in commercial applications. Thus, specific markers might be added to strains for testing in the greenhouse; these markers might not be used in field tests if better methods are developed during the preliminary tests. In addition, specific traits might be added or subtracted from the GEM during the first greenhouse tests. These might include debilitating the organism so as to decrease the chances of population explosion in case of accidental release, for example with sensitivity to temperatures seen in nature but not in the greenhouse.

E. *Points to consider*

There are several points to consider in microcosm experiments:

(a) Emphasis should be placed on replicability, both within and between laboratories. It is extremely important for decision-makers to know not only that in one set of experiments a given technique proved sensitive, but that different researchers in different laboratories working with the same and different experimental systems agreed. It is important to support research on the same techniques in more than one laboratory.

(b) The same microcosms used to study monitoring techniques should also be used for tests of environmental effects on the ecosystem. This will allow one to answer the question: Is the absence of a set of genes (according to our measuring techniques) a good predictor of no ecosystem effect? This is extremely important, and should perhaps be a type of project to be supported immediately. This is because monitoring is easier to do than assessing environmental impact. Organisms are now ready for field trials, yet it is not possible to design adequate tests to predict environmental effects. At this time the best we can do is simply assay for the survival of the organism and its genes. It is important to determine if this is sufficient, namely, if the absence of a set of genes by our techniques means we need not worry about adverse environmental effects.

The upshot of this is that research teams should be multi-disciplinary. This is likely to be difficult to accomplish because there are few such teams.

V. QUALITY ASSURANCE

A. Introduction

Quality assurance is concerned with the degree of confidence one has in a methodology. This involves both the sensitivity (probability of false negatives) and the specificity (probability of false positives) of the technique. All of the monitoring techniques mentioned in section III may give false negatives (the GEM is present but the technique does not detect it) and false positives (the technique gives a positive result even though the GEM is not present).

Prior to release into the environment, each monitoring technique must be calibrated in order to determine its ability to discriminate between the GEM and other microorganisms in the environment. There are two types of calibration experiments needed for each GEM, one to test for false positives and one for false negatives.

B. Testing for sensitivity and specificity

Microcosm experiments are necessary to test for sensitivity (false negatives). These involve determining the smallest populations that can reliably be detected with each technique as well as estimating precision and accuracy. For example, the sensitivity of a DNA probe should be calibrated by first determining whether it will hybridize to the rDNA and then determining the minimum density of a specific GEM which will be detected in a background of nonhomologous DNA.

Experimental conditions for a soil microorganism should range from pure culture studies in liquid medium, to pure culture studies in sterile soil, to experiments carried out in soil cores taken directly from the field with no pretreatment. One important aspect of these tests is that it should be possible to estimate population densities of the GEM accurately using another reliable technique. For example, in a pure liquid culture, optical density or microscopic counts can be used as a calibrating technique.

Specificity (false positives) must be studied using field samples. For example, testing immunofluorescence techniques involves observing stained and unstained samples from the environment for autofluorescence of organic material and specific or nonspecific binding of the antibody to other microorganisms (see section IIIB1 on immunological techniques).

Any attempt to use molecular probes for detection of GEM's first has to be calibrated with known rDNA and DNA from non-recombinant organisms. The probe should be tested to determine the amount of cross-hybridization interference due to common DNA sequences. The specificity of DNA probes is critically related to the stringency of the hybridization between the labelled probe and the tested DNA. By changing the temperature of the hybridization and wash conditions one can either increase or decrease the homology needed for heteroduplex formation. The monitoring protocol must be engineered so as to maximize sensitivity while minimizing cross-hybridization. Finally, specificity of DNA probes must be tested by assaying for homologous DNA in natural samples.

C. Testing for linkage between markers and rDNA

The ability of genetic markers to follow a GEM with a minimum of false postives and negatives critically depends on the degree of linkage between the marker gene and the rDNA. If they are not closely linked on the bacterial or plasmid chromosome, recombination will eventually destroy their association. This results in both a failure to recognize the GEM and misidentifying a non-recombinant organism as a GEM.

Therefore, laboratory studies must be done to determine the degree of linkage between the rDNA and the the marker. A crucial microcosm experiment involves monitoring known densities of the GEM with both the genetic marker and another monitoring technique such as colony hybridization that allows the investigator to detect the product of the rDNA or the rDNA itself. Such an experiment provides information on the stability of the marker gene and the rDNA.

D. Summary

In summary, all monitoring techniques require calibration tests to show that they can accurately identify the GEM in a background of non-recombinant organisms and to determine the minimal densities of the GEM that can be detected.

VI. CONCLUSIONS

A. Monitoring techniques

The techniques for monitoring GEM's discussed in section III can be ranked in a hierarchy in which cost increases with information received concerning the rDNA. Cost and sensitivity to low population densities are not correlated. Enrichment methods are likely to be more sensitive than plate counts at low densities; both are inexpensive. All molecular methods require isolation on plates, so sensitivity is a function of the proportion of GEM cells that grow and the selectivity of the medium.

Selectivity can be increased by adding markers for unusual growth characteristics or product formation, but the cost here is not in monitoring, but in modification of the GEM. Adding components to the selective medium entails minimal cost increase.

Thus, the techniques are (with numbers in increasing order of information received):

- (1a) selective plating
- (1b) selective enrichment
- (2a) fluorescent antibodies
- (2b) monoclonal antibodies
- (3) colony hybridization
- (4) restriction maps
- (5) restriction maps with radioactively labeled DNA probes
- (6) genomic sequencing

The suggested field monitoring protocol stresses qualitative, extensive sampling, concentrating on finding the few GEM's that might be transported over distances or those that might last a long time (see section II). Sampling might involve a succession of techniques, starting with the least expensive and proceeding to molecular techniques as simple methods begin to give negative results. The latter will occur either as time passes or as distance from the source increases.

For example, the gene for the _Bacillus thuringensis_ endotoxin has been added to a strain of _Pseudomonas fluorescens_ (see the second example below). The initial monitoring protocol might involve the use of a genomic marker, such as the ability to produce a chromogenic substance on a certain agar medium. If after awhile strains with the marker can no longer be found, then it might be necessary to begin to assay _Pseudomonas_ isolates with a DNA probe for the endotoxin gene, using colony hybridization techniques. Other species of _Pseudomonas_ should be tested in case there is gene transfer between them, and other genera known to be able to receive plasmids from _Pseudomonas_ should also be monitored for the rDNA sequence. Next, if colony hybridization produces only negative results, it may be appropriate to map some _Pseudomonas_ isolates with restriction enzymes to check the sensitivity of colony hybridization.

B. _Scenarios for protocol development_

Development of sampling protocols will require information on the properties of the specific GEM's to be released, combined with general knowledge of biological processes. The important issues for protocol development can be summarized in three questions:

(1) What are the possible adverse effects of release of the GEM? A list of these will suggest:

(1a) In what geographical locations and habitats should monitoring efforts be concentrated?

(1b) What species of plants and animals should be monitored for possible impacts?

(2) How fast and by what mechanisms does the GEM and its rDNA spread? Answers to this will suggest:

(2a) Where should monitoring efforts be concentrated geographically?

(2b) In what parts of the environment (habitats) should should efforts be concentrated?

(3) How might the GEM or its rDNA be modified, especially, how readily does the rDNA separate from plasmid and genomic markers? Answers to this will suggest: What combination of markers and direct molecular assays for rDNA should be used to most effectively follow the GEM and the rDNA?

Three examples of protocol development are discussed below. Some of the questions asked may already have been answered by the producers of the GEM's; nevertheless they are presented as examples of the types of issues that are important to regulatory agencies. The numbering and lettering follows the scheme just mentioned.

The first example is the "ice-minus" strain of *Pseudomonas syringae* studied by Lindow and colleagues (Lindow 1983) and developed by Microlife Technics, Inc. This strain is intended for introduction into crop fields to reduce frost damage. When strains of *Pseudomonas syringae* isolated from crop leaf surfaces are added to plant leaf surfaces in greenhouses, frost forms at relatively high temperatures (Lindow 1983). A strain has been found that when sprayed on leaf surfaces reduces frost damage (Lindow 1983). It probably acts by outcompeting the ice-nucleation-active (INA) strains for space or nutrents on the leaf surface (Cook and Baker 1983).

Strains of _Pseudomonas syringae_ are plant pathogens of a wide variety of species, including pear, cherry, wheat, lettuce, sugar beet, soybean, sunflower, rye, rice, barley and lilac (Krieg 1981). They affect plants in at least two ways; by toxin production (Daly and Deverall 1983) and by ice nucleation (Lindow 1983). Strains of _P. syringae_ are sometimes referred to as "pathovars" (abbreviated "pv."; Krieg 1981), and sometimes as separate species (Agrios 1978). The habitat of _P. syringae_ is the leaf surface; continued association with host plants is important to survival (Schroth et al. 1981). It is widely distributed; in a study of the distribution of _P. syringae_, strains were found on 48 of 59 plant species examined (in 23 of 27 families) in North America (Lindow et al. 1977). In this study, all samples were devoid of obvious symptoms of infection. However, the authors stated that the 800 isolates obtained may have included pathogens of different plants.

(1) _Possible adverse effects_.

Based on the information given above, potential adverse effects include: (i) Pathogenicity of the original strain used to produce the GEM towards valuable species such as crops or endangered species; (ii) accidental induction of or increase in pathogenicity towards valuable species during laboratory culture and manipulation; (iii) accidental decrease in pathogenicity towards undesirable species, that is, weeds or potential weeds.

These suggest that:

(1a) Monitoring for the GEM should concentrate on leaf surfaces, with less attention paid to soil. This is because the potential adverse effects occur on leaf surfaces and because survival is best on leaf surfaces. Ideally, all local species should be monitored, because it is always possible that a rare species is kept rare by pathogenic effects of _P. syringae_, and with a less virulent strain the species may become a weed.

However, it is not practical to test all species, so a decision must be made, based on feasibility of the test and likelihood of adverse effects. The latter might be judged from general knowledge of the biology of local plant species.

(1b) All local plant species should be monitored for gross changes in population density and visible indicators of disease. Monitoring should concentrate on crops, endangered species and species judged likely to be weeds.

The above suggestions for monitoring can be made more effective by a series of laboratory and greenhouse tests: (A) The growth and survival characteristics of the GEM can be tested on local species of plants. If survival is greater on certain species, these can be emphasized in the field. (B) Experiments with the GEM and closely related pathogenic strains can be used to discover which species of plants are most likely to show obvious signs of disease. (C) Survival and growth in the soil can be tested using the specific strain designated for release.

(2) Spread

The rate of movement of P. syringae in the soil is very amenable to laboratory and greenhouse tests. Mechanisms can be studied by varying environmental conditions, for example adding or subtracting earthworms, arthropods or host plants. The effects of fertilization and of greenhouse airflow patterns on migration can be studied. In general, these and other factors should be varied so as to give the best possible opportunities for the particular factor to have an effect. This hopefully gives an overestimate of the importance of each factor, which allows one to develop worst-case scenarios.

(2a) Monitoring in the field should extend beyond the limits of spread calculated in based on greenhouse experiments. Occasionally distant populations of disease-sensitive plants or species on which P. syringae is often found should be monitored for rare long-distance dispersal.

(2b) Monitoring should concentrate on the leaf surfaces found most likely to give positive results in the greenhouse.

(3) **GEM or rDNA modification**

The ice-minus property of the strains may be due either to deletions of the nucleating proteins or to modifications of them. This difference is crucial, because modified proteins may recombine or mutate back to nucleating forms, whereas deletions are very unlikely to return to wild-type function. In addition, if a natural protein is only modified, then monoclonal antibodies for the protein or molecular probes for the rDNA can be used.

If the rDNA is simply modified, then it may be modified further in the field; this may cause molecular techniques to miss rDNA and may cause reversion to ice nucleating activity. Modification of rDNA in the field can be tested in the laboratory and greenhouse by inoculating plants with the GEM and following it through time with several different molecular assays. For example, colony hybridization and restriction mapping with and without probes can be used to measure rDNA density over time.

The molecular method giving the greatest apparent recovery is not necessarily the best one, because it could give false positives. This can be tested in the laboratory by constructing several strains with similar but not identical rDNA and observing how the probe results correlate with ice nucleating activity.

Laboratory and greenhouse experiments can be used to compare the power and selectivity of different monitoring methods. Selective plating and enrichment can be used to follow markers; fluorescent antibodies against surface proteins can be used to follow the GEM; and molecular techniques can be used to follow the rDNA. The rates of die-off of the GEM measured by these different techniques will indicate the relative efficiencies of the methods.

In addition, these experiments can be used to test for separation of the ice-nucleation inactivity (INI) property from plasmid and chromosomal markers. This includes both loss of markers and transfer of the INI property to other strains of P. syringae and other species of bacteria. For example, greenhouse pots can be set up with host plants and several INI strains with different markers. The pots are then sampled over time, and P. syringae strains isolated and tested for the presence of the markers and INI. This will indicate the rate of die-off of the strains and whether the markers separate from the INI property. The experiments can be repeated, adding INA strains with distinguishable markers. Transfer of the INI property can thus be assayed.

These experiments will indicate how tightly linked are the INI property, the rDNA and the markers. The results will suggest which combination of molecular, immunological and standard microbiological methods follows the INI property and the GEM most efficiently.

The second example concerns the strain of Pseudomonas fluorescens that contains the gene for the Bacillus thuringensis endotoxin (Bioscience News Update 1985). The aim of this introduction is to poison the black cutworm, a parasite of corn roots, much as gypsy moth larvae are poisoned by B. thuringensis (Doane and McManus 1981). When B. thuringensis (hereafter B.t.) sporulates, it produces a parasporal crystal, a protein crystal inside the cell but outside the spore. The mixture of spores and crystals is sprayed onto leaves and the caterpillars consume them in the process of consuming leaves. The crystal dissolves in the alkaline gut of Lepidoptera, releasing toxic proteins (Deacon 1983). These act to poison the Na,K-ATPase in the gut wall and eventually lead to the disintegration of the gut wall (Dr. Leigh English, pers. comm., Harvard University). The spores invade the body chamber, germinate and grow. Feeding ceases soon after ingestion of spores and crystals; death is usually caused by the protein and sometimes by bacterial growth.

(1) *Possible adverse effects*

(i) If the added strain of *P. fluorescens* is transported to nearby areas, it might have the same effect on Lepidoptera in those areas. If, for example, a nearby grassland has an endangered species of butterfly, the transport of the gene for the *B. thuringensis* endotoxin into the area might significantly affect the survival of the species. (ii) *B. thuringensis* endotoxin kills Lepidopteran larvae by poisoning their Na,K-ATPase in the alkaline environment of their gut. Other orders of insects are usually not sensitive to this strain of *Bacillus* (Deacon 1983). However, it may be possible for the toxin to affect other important bacteriovores, for example protozoa and nematodes. If the latter are as important as they seem to be in nutrient recycling in the rhizosphere (Elliott et al. 1979), then poisoning microbial food webs might have significant effects on the entire soil ecosystem, including plants. (iii) *P. fluorescens* is known to be an animal pathogen (Bergan 1981).

(1a) Thus, development of an effective monitoring protocol should involve answering such questions as are suggested by this scenario: Are there other species nearby that might be adversely affected by transport of the GEM? This includes effects of the B.t. toxin as well as pathogenic effects of *P. fluorescens* itself on animals. If so, it may be necessary to monitor for what will probably be low densities of the GEM in the habitat of the endangered species.

(1b) In addition, it may be necessary to do qualitative population studies on the endangered species itself. Field monitoring might also include estimation of population sizes of bacteriovorous species likely to be affected. A relatively easy way to do this might involve (1) counting total densities of major groups, such as protozoa and nematodes, and (2) counting certain species individually. The latter can be chosen on the basis of ease of recognition or ecological importance (for example plant pathogenicity).

The value of such a sampling strategy must be explored in laboratory and greenhouse experiments. These include laboratory tests in which the GEM is fed to a variety of soil bacteriovores and if possible to other species of Lepidoptera.

(2) Spread

(2a) As above, monitoring in the field should extend beyond the limits of spread calculated based on greenhouse experiments.

(2b) *P. fluorescens* is a soil microorganism, so monitoring should concentrate in the soil. Greenhouse experiments should be done to determine whether recovery is better in the rhizosphere or in bulk soil, and at what soil depth.

(3) GEM or rDNA modification

Here, rDNA has obviously been added to the bacterium. Laboratory experiments can be done with modified B.t. proteins to determine how similar they must be to be picked up by the various molecular techniques. As above, combinations of molecular, standard microbiological and immunofluorescence techniques can be tested in the laboratory and greenhouse to observe how tightly linked these traits are in the GEM, and so to develop an effective, economical monitoring protocol.

The third example is the 2,4,5-T-degrading bacterium developed by Kellogg et al. (1981) using plasmid-assisted molecular breeding. This organism contains a combination of genes, each of which was already present in other genetic backgrounds. The mixing of genomes was performed by combining bacterial isolates, each with different combinations of plasmid-encoded pathways for degradation of various chlorinated carbon compounds. This was done in a chemostat with 2,4,5-T and other chlorinated compounds as energy sources. After running for 8 to 10 months, bacteria that were able to use 2,4,5-T as sole carbon source were isolated.

(1) Possible adverse effects

This new combination of genes may degrade important natural organic compounds and thereby cause environmental damage, for example by poisoning a plant species that contains a chlorinated organic compound, or by changing its rate of decomposition. In a recent review and listing of organic compounds found in plants, Robinson (1983) listed only three chlorinated molecules. In addition, these degradative pathways are fairly specific for certain kinds of molecules, so that the ability to use a very different chlorinated organic seems unlikely. Thus, the possibility that this new degradative ability would extend to a natural organic seems low.

The probability of such an effect occurring can be explored in greenhouse tests. In such an experiment, a dense suspension of the GEM is combined with other closely related strains in pots with plants known to contain a chlorinated organic compound. This design has the effect of also testing for the creation of new combinations of pathways by gene transfer in the pot. In the control pots, equally dense suspensions of the GEM cured of the degradative plasmids are inoculated along with the same closely related strains. Differences between treatment and control pots in plant health, growth rate and physiological parameters are measured.

The same type of experiment can be used to determine if the GEM might affect the rate of decomposition of plant material derived from species with chlorinated compounds. These experiments can be done in soil and aqueous microcosms.

Finally, *P. cepacia*, like *P. fluorescens*, is an animal pathogen (Bergan 1981). This can be tested in single-species laboratory toxicity tests.

Positive results would suggest that:

(1a) Monitoring should concentrate near or on plant species that contain chlorinated compounds. These might be found in references such as Robinson (1983). Monitoring should include dead plant material. If animal pathogenicity is deemed potentially important in the strain due to be released, then monitoring should also include possible animal hosts.

(1b) Local plant species with chlorinated compounds and potential local animal hosts should be monitored for unusual changes in appearance or population density.

(2) Spread

Rates of dispersal can be studied as described in the first two examples.

GEM or rDNA modification

Analysis of rDNA would be difficult because the components of the degradative plasmids are all or mostly all found in nature. Given this, along with the fact that chlorinated compounds are very rare in nature, it might be concluded that assaying for the rDNA in this organism might not be worthwhile. However, if a certain portion of the rDNA is not found naturally with the use of DNA probes, then the appropriate probe can be used to follow this portion.

This suggests that molecular assays may not be very useful in this case. For this organism, the easiest and most direct assay is the ability to grow on plates or in enrichment cultures with 2,4,5-T as sole carbon source (Kilbane *et al.* 1983).

C. Research needs

The primary research need is for microcosm tests of monitoring protocols. This should involve testing specific sets of techniques with representative organisms, preferably GEM's due to be released. With the exception of genomic sequencing, all of the techniques mentioned in section III are well-established laboratory procedures. Thus, research support is needed for their application, not their development. Research should stress comparative studies of the techniques and studies of their application to natural systems. Long-term research on microcosm design for assessing fate and environmental impacts is necessary either to bring us closer to the goal of a standardized set of microcosms with standardized measurements, or to ascertain that such standardization is not possible given natural variability. Replicability between laboratories is an important issue for EPA to consider in supporting research; it may be worthwhile to support similar projects in more than one laboratory.

Microcosm tests should include those mentioned in section V on quality assurance. These involve testing microcosm populations of GEM's for sensitivity of the techniques. In addition, laboratory and environmental samples must be tested for false positive results.

Two important aspects of the application of these techniques to natural situations are: (1) Partitioning of GEM's into different parts of the environment. This can be explored experimentally with GEM's which are to be released or with very similar strains. The purpose is to see if GEM's survive or grow better in different habitats, for example different depths of the soil, water column or aquatic sediments. This information might be useful in designing sampling strategies for low-density populations.

(2) Desorption of cells from surfaces for subsequent plating and molecular analysis. There is already much literature, so the first step should be the compilation of existing techniques. The second step should then be to explore the different methods experimentally.

The research needs described here can be divided up into short- and long-term efforts. Short-term efforts involve testing existing microcosm systems, monitoring techniques and desorption methods. The purpose of this is to establish an interim set of protocols. Much information can be gathered in a matter of one or two years, using GEM's already prepared for release.

Three organisms that may be useful for short-term research are: (1) the ice-minus *Pseudomonas syringae* with a gene deletion; (2) the *P. fluorescens* with a single gene for the *B. thuringensis* endotoxin added; and (3) the 2,4,5-T degrading strain of *P. cepacia* with several added genes. These three are quite different in their ecological effects and in the nature of the genetic manipulation, so they form a good starting point. Consideration should also be given to other types of organisms planned for release.

Long-term projects involve microcosm design for simultaneous assessment of monitoring techniques, fate and effects, new techniques for desorption, new marking methods, and further studies on how GEM's partition themselves in different parts of the environment. Advances in molecular techniques for examining changes in genomic structure should be followed and applied as appropriate to monitoring protocols.

VI. LITERATURE CITED

1. Agrios, G.N. 1978. Plant Pathology, second edition. New York, Academic Press.
2. Anderson, S.S. 1975. Plasmid transfer in *E. coli*. J. Infec. Diseases 175: 686-687.
3. Atlas, R.M. and R. Bartha. 1981. Microbial Ecology: Fundamentals and Applications. Reading, Mass., Addison-Wesley Publishing Company.
4. Bergan, T. 1981. Human- and animal-pathogenic members of the genus *Pseudomonas*. In: The Procaryotes, Volume 1, edited by M.P. Starr *et al*. Chapter 59. New York, Springer-Verlag.
5. Bioscience News Update. Gene-splicing field trials schedule. Bioscience 35: 147.
6. Bohlool, B.B. and E.L. Schmidt. 1980. The immunofluorescence approach in microbial ecology. Adv. Microb. Ecol. 4:203-242.
7. Chilton, M.D., A. DeFramond, R. Fraley, *et al*. 1983. Ti and Ri plasmids as vectors for genetic engineering of higher plants. Abstract, 15th Miami Winter Symposium, Miami, January 17-21, 1983.
8. Church, G.M. and W. Gilbert. 1984. Genomic sequencing. Proc. Nat. Acad. Sci. USA. 81: 1991-1995.
9. Cook, R.J. and K.F. Baker. 1983. The Nature and Practice of Biological Control of Plant Pathogens. St. Paul, Minnesota, The American Phytopathological Society.
10. Daly, J.M. and B.J. Deverall, eds. 1983. Toxins and Plant Pathogenesis. New York, Academic Press.
11. Deacon, J.W. 1983. Microbial Control of Plant Pests and Diseases.

Washington, D.C., American Society for Microbiology.

12. Doane, C.C. and M.L. McManus, eds. 1981. The Gypsy Moth: Research Toward Integrated Pest Management. Washington, D.C., U.S. Department of Agriculture.

13. Draggan, S. and P. Van Voris. 1979. The role of microcosms in ecological research (An introduction to this special issue on microcosms). Int. J. Env. Studies 13: 83-85.

14. Elliott, E.T., C.V. Cole, D.C. Coleman, R.V. Anderson, H.W. Hunt and J.F. McClellan. 1979. Amoebal growth in soil microcosms: A model system of C, N, and P trophic dynamics. Int. J. Env. Studies 13: 169-174.

15. Fraley, R.T., R. Horsch and S.G. Rogers. 1983. University of California at Los Angeles - Keystone Meeting, Keystone, Colorado, March 1983.

16. Geisy, J.P., ed. 1980. Microcosms in Ecological Research. Washington, D.C., U.S. Department of Energy.

17. Genewatch. 1983. Ice-enhancing bacteria. Genewatch 1(3-4): 11.

18. Gillett, J.W. and J.M. Witt, eds. 1977. Terrestrial Microcosms. Proceedings of a workshop on terrestrial microcosms, 1977.

19. Grice, G.D. and M.R. Reeve, eds. 1982. Marine Mesocosms. New York, Springer-Verlag.

20. Grunstein, M. and D.S. Hogness. 1975. Colony hybridization: A method for the isolation of cloned DNAs that contain a specific gene. Proc. Nat. Acad. Sci. 72: 3961-3965.

21. Herrera-Estrella, L., A. Depicker, M. Van Montagu, et al. 1983. Expression of chimeric genes transfered into plant cells using a Ti-plasmid-derived vector. Nature 303: 209-213.

22. Hobbie, J.E., R.J. Daley and S. Jasper. 1977. Use of Nuclepore filters for counting bacteria by fluorescence microscopy. Appl. Envir. Microb. 33: 1225-1228.

23. Kellogg, S.T., D.K. Chatterjee and A.M. Chakrabarty. 1981. Plasmid-assisted molecular breeding: New technique for enhanced biodegradation of persistent toxic chemicals. Science 214: 1133-1135.

24. Kemp, J. 1983. UCLA - Keystone Meeting, Keystone, Colorado, March, 1983.

25. Kilbane J.J., D.K. Chatterjee, and A.M. Chakrabarty. 1983. Detoxification of 2,4,5-trichlorophenoxyacetic acid from contaminated soil by _Pseudomonas cepacia_. Appl. Envir. Microb. 45: 1697-1700.

26. Kimball, K.D. and S.A. Levin. 1985. Limitations of laboratory bioassays: the need for ecosystem-level testing. Bioscience 35: 165-171.

27. Kinne, O. 1977. International Helgoland Symposium "Ecosystem Research": Opening Address. Helg. wissen. Meeresunt. 30: 1-7.

28. Krieg, N.R., ed. 1981. Bergey's Manual of Systematic Bacteriology, Volume 1. Baltimore, Maryland, Williams and Wilkins.

29. Langer, P.R., A.A. Waldrop and D.C. Ward. 1981. Enzymatic synthesis of biotin-labeled polynucleotides: Novel nucleic acid affinity probes. Proc. Nat. Acad. Sci. 78: 6633-6637.

30. Levy, S.B. 1984. Survival of plasmids in _E. coli_. In Arber, W., K Illmensee, W.J. Peacock and P. Starling (eds.) Genetic Manipulation. Impact on Man and Society. Cambridge, Cambridge University Press.

31. Levy, S.B., B. Marshall, A. Orderenk, and D. Rouse-Eagle. 1980. Survival of _Escherichia coli_ host-vector systems in the mammalian intestine. Science 209: 391-394.

32. Lewin, R. 1983. Genetic probes become ever sharper. Science 221: 1167.

33. Lindow, S.E. 1983. Methods of preventing frost injury caused by epiphytic ice-nucleation-active bacteria. Plant Disease 67: 327-333.

34. Lindow, S.E., D.C. Arny and C.D. Upper. 1977. Distribution of epiphytic

ice-nucleation-active strains of *Pseudomonas syringae*. Proc. Amer. Phytopath. Soc. 4:107.

35. Lynch, J.M. and N.J. Poole, eds. 1979. Microbial Ecology: A Conceptual Approach. Oxford, Blackwell Scientific Publications.
36. Maniatis, T., E.F. Fritsch and J. Sanbrook. 1982. Molecular Cloning: a Laboratory Manual. Cold Spring, New York, Cold Spring Harbor Laboratory.
37. Marshall, K.C. 1976. Interfaces in Microbial Ecology. Cambridge, Mass., Harvard University Press.
38. Masui, Y., T. Mizuno and M. Inouye. 1984. Novel high-level expression cloning vehicles: 10^4-fold amplification of *Escherichia coli* minor protein. Biotech. 2: 81-85.
39. Maxam, A.M. and W. Gilbert. 1980. Sequencing end-labeled DNA with base-specific chemical cleavages. Methods Enzymol. 65: 499-560.
40. Meyers, J.A., D. Sanchez, L.P. Elwell and S. Falkow. 1976. Simple agarose gel electrophoresis method for the identification and characterization of plasmid deoxyribonucleic acid. J. Bact. 127: 1529-1537.
41. Peacock, W.J. 1983. Gene transfer in agricultural plants. Abstract, 15th Miami Winter Symposiuma, Miami, January 17-21, 1983.
42. Porter, K.G. and Y.S. Feig. 1980. The use of DAPI for identifying and counting aquatic microflora. Limnol. Oceanog. 25: 943-948.
43. Pritchard, P.H. and A.W. Bourquin. 1984. The use of micrcosms for evaluation of interactions between pollutants and microorganisms. Adv. Microb. Ecol. 7: 133-216.
44. Robinson, T. 1983. The Organic Constituents of Higher Plants. North Amherst, Massachusetts, Cordus Press.
45. Russek, E. and R.R. Colwell. 1983. Computation of most probable numbers. Appl. Env. Microb. 45: 1646-1650.

46. Sagik, B.P. and C.A. Sorber. 1979. The survival of H-V systems in domestic sewage treatment plants. Recomb. DNA Tech. Bull. 2: 55-61.
47. Sanger, F. et al. 1977. Proc. Nat. Acad. Sci. USA. 74: 5463.
48. Sayler, G.S., M.S. Shields, E.T. Tedford, A. Breen, S.W. Hooper, K.M. Sirotkin, and J.W. Davis. 1985. Application of DNA-DNA colony hybridization to the detection of catabolic genotypes in environmental samples. Appl. Envir. Microb. 49: 1295-1303.
49. Schell, J., M. Van Montagu, J.P. Hernalsteens, et al. 1983. Ti plasmids as experimental gene vectors for plants. Abstract, 15th Miami Winter Symposium, Miami, January 17-21, 1983.
50. Schiff, W. and W. Klingmuller. 1983. Experiments with Escherichia coli on the dispersal of plasmids in environmental samples. Recomb. DNA Tech. Bull. 6: 101-102.
51. Schroth, M.N., D.C. Hildebrand and M.P. Starr. 1981. Phytopathogenic members of the genus Pseudomonas. In: The Procaryotes, Volume 1, edited by M.P. Starr et al. Chapter 60. New York, Springer-Verlag.
52. Sibley, C.G. and J.E. Alquist. 1983. Phylogeny and classification of birds based on the data of DNA-DNA hybridization. Current Ornithology 1: 245-292.
53. Sieburth, J.M. 1979. Sea Microbes. New York, Oxford University Press.
54. Southern, E.M. 1975. Detection of specific sequences among DNA fragments separated by gel electrophoresis. J. Mol. Biol. 98: 503-517.
55. Szwerinski, H., S. Gaiser and D. Bardtke. 1985. Immunofluorescence for the quantitative determination of nitrifying bacteria: interference of the test in biofilm reactors. Appl. Microbiol. Biotechnol. 21: 125-128.
56. Watson, J. 1980. Foreword to the special issue on the Experimental Lakes Area. Can. J. Fish. Aq. Sci. 37: 311-312.
57. Witt, J.M. and J.W. Gillett, eds. 1979. Terrestrial Microcosms and Environmental Chemistry. Proceedings of two colloquia, held June 13-14, 1977, at Oregon State University, Corvallis, Oregon.

APPENDIX
LIST OF PARTICIPANTS

AAAS/EPA Biotechnology Workshop
29 April — May, 1984

Workshop Chairman

    Gilbert S. Omenn, Dean
    School of Public Health and Community Medicine
    University of Washington
    Seattle, WA

Health Effects Review Group

    D. Michael Gill (Chairman)
    Department of Molecular Biology and Microbiology
    Tufts University Medical School
    Boston, MA

    Susan Gottesmann
    Laboratory of Molecular Biology
    National Cancer Institute
    Bethesda, MD

    Dennis J. Kopecko
    Department of Bacterial Immunology
    Walter Reed Army Institute of Research
    Washington, DC

    Richard Novick
    Department of Plasmid Biology
    Public Health Research Institute
    New York, NY

    Mark Townsend (Rapporteur)
    Office of Toxic Substances, EPA
    Washington, DC

    David Kleffman (Author)
    Office of Research and Development, EPA
    Washington, DC

    Clinton Kawanishi
    Health Effects Research Laboratory, EPA
    Research Triangle Park, NC

    Michael Waters
    Health Effects Research Laboratory, EPA
    Research Triangle Park, NC

    Daphne Kamely
    Office of Research and Development, EPA
    Washington, DC

Mark Segal
Office of Toxic Substances, EPA
Washington, DC

Stephen L. Johnson
Office of Pesticide Programs, EPA
Washington, DC

Tina Levine
Office of Toxic Substances, EPA
Washington, DC

William Schneider
Office of Pesticide Programs, EPA
Washington, DC

Irving Mauer
Office of Pesticide Programs, EPA
Washington, DC

Environmental Effects Review Group

R. Darryl Banks
New York State Department of
  Environmental Conservation
Albany, NY

Philip Regal
Department of Ecology and Behavioral Biology
University of Minnesota
Minneapolis, MN

Frances Sharples (Chairman)
Office of Institutional Planning
Oak Ridge National Laboratory
Oak Ridge, TN

Guenther Stotzky
Department of Biology
New York University
New York, NY

Frederick Betz (Rapporteur)
Office of Pesticide Programs, EPA
Washington, DC

Charles Hendricks (Author)
Office of Research and Development, EPA
Washington, DC

Al Bourquin (Author)
Environmental Research Laboratory, EPA
Sabine Island
Gulf Breeze, FL

Gary S. Saylor
Graduate Program in Ecology
Department of Microbiology
University of Tennessee
Knoxville, TN

William G. Wells (Rapporteur)
Office of Public Sector Programs, AAAS

Charles Plost
Office of Research and Development, EPA
Washington, DC

Christon J. Hurst (Author)
Environmental Monitoring Systems Laboratory, EPA
Cincinnati, OH

John A. Santolucito
Environmental Monitoring Systems Laboratory, EPA
Las Vegas, NV

Nancy Chiu
Office of Toxic Substances, EPA
Washington, DC

Michael Callahan
Office of Toxic Substances, EPA
Washington, DC

Andrew Jovanovich
Office of Pesticide Programs, EPA
Washington, DC

Amy S. Rispin
Office of Pesticide Programs, EPA
Washington, DC

Michael Dellarco
Office of Research and Development, EPA
Washington, DC

Containment and Control Technologies Review Group

Aileen Compton
Smith, Kline and French Laboratories, Inc.
Philadelphia, PA

Charles Cooney (Chairman)
Department of Chemical Engineering
MIT
Cambridge, MA

Hester Kobayashi
Standard Oil of Ohio Research Center
Cleveland, OH

Seth Pauker
Biogen Research Corporation
Cambridge, MA

Barry D. Gold (Rapporteur)
Office of Public Sector Programs, AAAS

N. Dean Smith
Industrial Environmental Research Laboratory, EPA
Research Triangle Park, NC

John Burckle (Author)
Industrial Environmental Research Laboratory, EPA
Cincinnati, OH

Albert D. Venosa (Author)
Environmental Research Laboratory, EPA
Cincinnati, OH

Bala Krishnan
Office of Research and Development, EPA
Washington, DC

Elizabeth Ward
Office of Toxic Substances, EPA
Washington, DC

Larry Longanecker
Office of Toxic Substances, EPA
Washington, DC

Organizers and Staff

John R. Fowle, III
Office of Research and Development, EPA
Washington, DC

Anne Hollander
Office of Toxic Substances, EPA
Washington, DC.

Morris A. Levin
House Committee on Science & Technology
Washington, DC

Albert H. Teich, Project Director
Office of Public Sector Programs, AAAS

Jill H. Pace, Project Associate
Office of Public Sector Programs, AAAS

Mary I. Haddock, Administrative Secretary
Office of Public Sector Programs, AAAS

Special Guests*

    Carl Gerber
    Office of Research and Development, EPA
    Washington, DC

    William Waugh
    Office of Toxic Substances, EPA
    Washington, DC

    Carl Mazza
    Office of Toxic Substances, EPA
    Washington, DC

    Donald Clay
    Office of Toxic Substances, EPA
    Washington, DC

    Bernard Goldstein
    Office of Research and Development, EPA
    Washington, DC

    Scott Baker
    Office of Research and Development, EPA
    Washington, DC

*Opening Session

Other Noyes Publications

BIOTECHNOLOGY IN FOOD PROCESSING

Edited by

Susan K. Harlander and Theodore P. Labuza

University of Minnesota
Department of Food Science and Nutrition

"...I am confident that efforts to optimize biotechnical organisms for hyperproduction will uncover new principles..."—from the Foreword by *Nobel Laureate Joshua Lederberg, President, Rockefeller University.*

This important, excellent and comprehensive book reviews the role, applications, technology, bioprocess engineering and economics of biotechnology in food processing.

Biotechnology will have a dramatic effect on the food processing industry, creating exciting opportunities in new product development and differentiation, cost reduction, and creation of novel processing methods. The challenge for food processors will be to determine how, when, and in what areas biotechnology will impact their industry.

Written by a distinguished group of contributors with a foreword by Nobel Laureate, Joshua Lederberg, one of the key figures in the development of biotechnology, the book will be the base line from which the application of biotechnology to food processing will reach the next level of development.

CONTENTS

1. **Biotechnology: Its Potential Impact on Traditional Food Processing**
   R. Moshy
2. **Regulatory Issues in the Food Biotechnology Arena**
   S. McNamara
3. **Federal Regulation of Food Biotechnology**
   S. Ronk
4. **Enzymology and Food Processing**
   S. Neidleman
5. **Protein Engineering: Potential Applications in Food Processing**
   R. Wetzel
6. **Biopolymers and Modified Polysaccharides**
   A. Sinskey, S. Jamas,
   D. Easson Jr., C. Rha
7. **Use of Microorganisms in the Production of Unique Ingredients**
   N. Trivedi
8. **Potential Application of Plant Cell Culture**
   D. Evans, W. Sharp
9. **Application of Genetic Engineering Techniques for Dairy Starter Culture Improvement**
   L. McKay
10. **Production of L-Ascorbic Acid from Whey**
    T. Cayle, J. Roland, D. Mehnert,
    R. Dinwoodie, R. Larson,
    J. Mathers, M. Raines, W. Alm,
    S. Ma'ayeh, S. Kiang, R. Saunders
11. **The Genetic Modification of Brewer's Yeast and Other Industrial Yeast Strains**
    I. Russell, R. Jones, G. Stewart
12. **Lactobacilli in Food Fermentations**
    B. Chassy
13. **Food Fermentation with Molds**
    R. Buchanan
14. **Unitization of Fermented Foods: An Application of Fermentation Technology**
    M. Sfat
15. **Separation Technology for Bioprocesses**
    R. Grabner
16. **Scale Up of a Fermentation Process**
    P. Senior
17. **The Use of Enzymes for Waste Management in the Food Industry**
    S. Shoemaker
18. **Biosensors for Biological Monitoring**
    R. Fitts
19. **Strategies for Commercialization of Biotechnology in the Food Industry**
    D. Wheat
20. **Cost Reductions in Food Processing Using Biotechnology**
    D. Jackson
21. **Profit Opportunities in Biotechnology for the Food Processing Industry**
    N. Newell, S. Gordon

ISBN 0-8155-1073-X (1986)

323 pages

Other Noyes Publications

ORGAN FUNCTION TESTS IN TOXICITY EVALUATION

Edited by

Charles A. Tyson
SRI International

Daljit S. Sawhney
Office of Pesticides and Toxic Substances
U.S. Environmental Protection Agency

A survey of biochemical and physiological tests designed to detect and assess organ function for toxicity evaluation—particularly in the hepatic, renal, cardiovascular, and pulmonary systems—is presented in this book. Scientific bases for, advantages of, and limitations to these tests are provided. In addition, procedures and tests needing further development and validation are identified.

The emphasis here is on the identification and description of tests or methods which can be used in conjunction with more traditional toxicity studies to offer increased sensitivity or predictivity over existing tests and provide more useful information to evaluate organ functions. Presently used tests and methodologies are discussed from the standpoint of their ability to detect pathophysiologic changes due to chemical exposure.

The review addresses the scientific merits of a particular test or technique, species sensitivity, cost-effectiveness, level of expertise required to conduct the test, its significance in assessing health hazards, and approaches that might improve the sensitivity of the test, if these are evident.

A condensed table of contents listing **chapter titles, authors, and selected subtitles** is given below.

CONTENTS

1. **RATIONALE AND USE OF FUNCTION TESTS IN TOXICITY TESTING—A REVIEW**
   W.H. Farland, C.A. Tyson, D.S. Sawhney
   Rationale for Use of Organ Function Tests
   Previous Reviews by Scientific Panels
   Other Reviews and Considerations

2. **TRADITIONALLY USED INDICATORS IN RELATION TO PATHOLOGIC LESIONS**
   C.A. Tyson, D.L. Story, C.E. Green, E.F. Meierhenry
   Biochemical Indicators in Relation to Pathology
   Indicator Changes and Their Possible Interpretation

3. **TESTS FOR HEPATOBILIARY FUNCTION**
   C. Mitoma
   Organ Function Tests
   Clinical Chemistry

4. **TESTS FOR RENAL FUNCTION**
   C. Mitoma
   Organ Function Tests
   Clinical Chemistry

5. **TESTS FOR PULMONARY FUNCTION**
   L.T. Juhos, D.P. Sasmore, D.G. Falconer, C. Tyson
   Biochemical Indicators
   Respiratory Function Tests
   Sonic and Radiographic Techniques
   Pathologic Evaluation

6. **TESTS FOR CARDIOVASCULAR FUNCTION**
   R.A. Jensen
   Arrhythmogenesis Screening
   Chemically Induced Cardiomyopathies
   Hemodynamic Evaluations
   Noninvasive Ventricular Function Evaluation
   Serum Enzymes

7. **SUMMARY, RECOMMENDATIONS, AND RESEARCH NEEDS**
   D.S. Sawhney, W.H. Farland, C.A. Tyson
   Pulmonary
   Cardiovascular
   Hepatic/Biliary Tract
   Renal

BIBLIOGRAPHY

ISBN 0-8155-1036-5 (1985)   237 pages

Other Noyes Publications

HANDBOOK OF CARCINOGEN TESTING

Edited by

Harry A. Milman
Office of Toxic Substances
U.S. Environmental Protection Agency

Elizabeth K. Weisberger
National Cancer Institute
National Institutes of Health

This definitive handbook of carcinogen testing is the first to outline in detail **all** methods, short-term **and** long-term, of identifying chemical carcinogens. It affords a total view of a bioassay, from its initial phases to its application, while at the same time defining its attributes and limitations. The book, presented in ten parts, is authored by a group of more than sixty international authorities on the subject of carcinogen testing.

CONTENTS

I. Predicting Carcinogenicity of Chemicals from Their Structure
1. Structural and Functional Criteria for Suspecting Chemical Compounds of Carcinogenic Activity—State-of-the-Art of Predictive Formalism

II. Epidemiological Investigations
2. Role of Epidemiology in Identifying Chemical Carcinogens

III. *In-Vitro* Tests
3. Overview of *In Vitro* tests for Genotoxic Agents
4. The *Salmonella* Mutagenicity Assay for Identification of Presumptive Carcinogens
5. Detection of Carcinogens Based on *In Vitro* Mammaliam Cytogenic Tests
6. Methods and Modifications of the Hepatocyte Primary Culture/DNA Repair Test
7. Cell Transformation Assays

IV. Limited Bioassays
8. Rat Liver Foci Assay
9. Lung Tumors in Strain A Mice as a Bioassay for Carcinogenicity
10. Tumorigenesis of the Rat Mammary Gland
11. SENCAR Mouse Skin Tumorigenesis

V. Long-Term Animal Bioassays
12. Design of a Long-Term Animal Bioassay for Carcinogenicity
13. Conduct of Long-Term Animal Bioassays
14. Selection and Use of the B6C3F1 Mouse and F344 Rat in Long-Term Bioassays for Carcinogenicity
15. Adequacy of Syrian Hamsters for Long-Term Animal Bioassays
16. Quality Assurance in Pathology for Rodent Carcinogenicity Studies
17. Statistical Evaluation of Long-Term Animal Bioassays
18. Considerations in the Evaluation and Interpretation of Long-Term Animal Bioassays for Carcinogenicity

VI. Bioassays for Insoluble Materials
19. Bioassays for Asbestos and Other Solid Materials

VII. Assays with Potential Utility
20. *In Vitro* Assay to Detect Inhibitors of Intercellular Cooperation
21. Alpha-Fetoprotein: A Marker for Exposure to Chemical Hepatocarcinogens in Rats
22. Ornithine Decarboxylase as a Marker of Carcinogenesis
23. Assay for Hepatic Peroxisome Proliferation to Select a Novel Class of Non-Mutagenic Hepatocarcinogens

VIII. Risk Estimation
24. Examination of Risk Estimation Models
25. Risk Assessment—Biological Considerations

IX. Regulatory Implications
26. Regulatory Implications—Perspective of the U.S. Environmental Protection Agency
27. New Approaches to the Regulation of Carcinogens in Food—The Food and Drug Administration
28. Workplace Carcinogens—Regulatory Implications of Investigations
29. Evaluation of Carcinogens—Perspective of the Consumer Product Safety Commission
30. International Aspects of Testing for Carcinogenicity and Regulation—A Selected Bibliography

X. Industry Perspective
31. An Industrial Perspective on Testing for Carcinogenicity

Index

ISBN 0-8155-1035-7 (1985)

638 pages

Other Noyes Publications

HANDBOOK OF MONOCLONAL ANTIBODIES
Applications in Biology and Medicine

Edited by

Soldano Ferrone
New York Medical College
Valhalla, New York

Manfred P. Dierich
University of Innsbruck
Innsbruck, Austria

Biological and medical applications of monoclonal antibodies are described in detail in this comprehensive handbook prepared by an outstanding panel of internationally known researchers.

Monoclonal antibodies are considered to be one of the most remarkable research achievements of recent times. Their practical applications already have had enormous impacts in many areas of medicine. The extensive utilization of these reagents in both *in-vitro* and *in-vivo* studies is covered in the book. Each chapter contains up-to-date and, often, previously unpublished data; thus making the book a useful tool for a wide range of professionals.

CONTENTS

1. Monoclonal Antibody Production—Principles and Practice
   S. Fazekas de St.Groth
2. Lymphocyte Subpopulations in Rodents—Functional Subpopulations in Rodents
   M.R. Greenwood, R.M. Parkhouse
3. Lymphocyte Subpopulation Separation by Monoclonal Antibodies
   R. Braun
4. Immunofluorescence Analysis of Leucocyte/Lymphocyte Membrane Antigens with Monoclonal Antibodies
   W. Ax, S. Schottler
5. Phenotyping of Human NK Cells with Monoclonal Antibodies
   D. Kraft, H. Rumpold, O. Eremin
6. Monoclonal Antibodies and Human Myelomonocytic Cells
   W. Knapp
7. Monoclonal Antibodies as Probes to Define Human T Cell Surface Structures and Functions
   S.C. Meuer, S.F. Schlossman, E.L. Reinherz
8. Monoclonal Antibodies as Tools to Study Receptor-Transducer-Effector Systems
   M.Schmitt, R.G. Painter, C.G. Cochrane
9. Monoclonal Antibodies in Enzymology
   F. Celada, B. Rotman
10. Monoclonal Antibodies in the Analysis of Structure and Function of Complement Components
    R. Burger
11. Monoclonal Antibodies and Coagulation
    K.G. Mann, J.A. Katzmann, W.B. Foster, D.N. Fass
12. Monoclonal Antibodies and Human Chorionic Gonadotropin—Mapping of Epitopes and Receptor-Interaction Sites
    S. Schwarz, R. Kofler, P. Berger, G. Wick
13. Monoclonal Antibodies and Viruses—Application to the Epstein-Barr System
    N. Mueller-Lantzsch
14. Monoclonal Antibodies and Herpes Simplex Virus Infections
    J.C. Glorioso, M. Levine
15. Monoclonal Antibodies and Bacteria
    T.F. Schulz, M.P. Dierich
16. Monoclonal Antibodies in Diagnostic Pathology
    S.E. Kabawat, F.I. Preffer, A.K. Bhan
17. Monoclonal Antibodies and Detection of Malignancies
    D.R. Schultz
18. Monoclonal Antibodies and Melanoma
    J.P. Johnson, G. Riethmuller
19. Monoclonal Antibodies and Characterization of Human Lymphoproliferative Malignancies
    D.M. Knowles
20. Monoclonal Antibodies and Human Histocompatibility Antigens
    P.G. Natali, A. Bigotti, C. Russo, K. Sakaguchi, M. Igarashi, S. Ferrone
21. Monoclonal Antibodies in the Therapy of Renal Transplant Rejection
    C.B. Carpenter
22. Monoclonal Antibodies as Therapeutic Tools in Medicine
    J.-F. Bach, L. Chatenoud
23. Monoclonal Antibodies in the Therapy of Malignant Diseases
    R.O. Dillman, I. Royston

Appendix—List of Commercially Available Monoclonal Antibodies

ISBN 0-8155-1034-9 (1985)

477 pages